U0930516

THE THIRD SPACE

CLUB, NIGHTCLUB & BAR

第三空间 会所、夜总会和酒吧

许雅杰 沈沉 译

高迪国际出版有限公司 编
大连理工大学出版社

图书在版编目 (CIP) 数据

第三空间：会所、夜总会和酒吧：英汉对照 / 高迪国际出版有限公司编；许雅杰，沈沉译. —大连：大连理工大学出版社，2013.10

ISBN 978-7-5611-8186-7

Ⅰ. ①第… Ⅱ. ①高… ②许… ③沈… Ⅲ. ①服务建筑-室内装饰设计-世界-图集②酒吧-室内装饰设计-世界-图集 Ⅳ. ① TU247-64

中国版本图书馆 CIP 数据核字 (2013) 第 206275 号

出版发行：大连理工大学出版社
（地址：大连市软件园路 80 号 邮编：116023）
印　　刷：上海锦良印刷厂
幅面尺寸：240mm × 290mm
印　　张：21
插　　页：4
出版时间：2013 年 10 月第 1 版
印刷时间：2013 年 10 月第 1 次印刷
策划编辑：袁　斌　刘　蓉
责任编辑：刘　蓉
责任校对：李　雪
封面设计：黄子平

ISBN 978-7-5611-8186-7
定　　价：328.00 元

电话：0411-84708842
传真：0411-84701466
邮购：0411-84703636
E-mail:designbookdutp@gmail.com
URL:http://www.dutp.cn

如有质量问题请联系出版中心：（0411）84709246 84709043

THE THIRD SPACE

CLUB, NIGHTCLUB & BAR

PREFACE Ⅰ 序言一

From sweaty basements to hi-tech entertainment temples – clubs have changed a lot from the beginnings to now. With the changes come new challenges. While club culture is a global phenomenon, the energy and atmosphere is created by the clubbers, which are rooted in their respective local cultures. This is also true for clubs in Asia, which often are under the pressure to survive in a transforming entertainment environment, often by integrating the needs of diverse audiences in one space. Unfortunately, too often the result is conceptual confusion. In order to avoid this, URBANTAINER had a look at the successful yet authentic European club scene.

The URBANTAINER team developed Club OCTAGON's space branding concept with these considerations in mind. While the overall concept is a reference to Europe's transformed post-industrial spaces, the design concept is inspired by the octagonal shape. With its prevalence in Korean traditional design, the designers and planners chose the octagon as design motif because of its specific representation of a human between the elements of land and heaven.

The motif runs throughout the club interior, the lighting, and it sets the structural rules for construction. The octagonal shape also allowed the designers to create the effect of a coliseum without the difficulties that come with building in a completely circular shape.

Whilst being optimal for viewing the stage from all angles and producing superb acoustics, the OCTAGON allowed for the presence of straight edges and thus better inclusion of technological equipment such as the Funktion-One surround sound system. In URBANTAINER's world, design is understood as a creative process of communication within a certain situation, rather than applying templates just because they work. Maintaining this mindset can be a challenge, but the results are always worth the extra effort. This is especially true for the realization of projects in the dynamic Megacities of Asia.

从闷热的地下室到高科技的娱乐圣殿——会所从一开始到现在已经改变了很多。变化的同时也带来了新的挑战。虽然会所文化已经成为一种全球现象，但是会所会员的活力和会所的气氛却植根于各自的文化之中。这点同样适用于亚洲的会所，这些会所常常顶着压力，通过整合一个空间中不同受众的需求来转变娱乐环境，从而求得生存发展。不幸的是，这样的装饰往往导致概念上的混乱。为了避免这种情况，URBANTAINER 的设计参考了颇有成效且正宗的欧洲会所场景。

考虑到上述因素，URBANTAINER 的团队开发了 OCTAGON 会所的空间品牌理念。虽然整体的概念参照了欧洲转型后的后工业化空间，但设计概念的灵感却来自于八角形的形状。随着八角形在韩国传统设计中的广泛应用，设计者和规划者选择将八角形作为设计主题，因为八角形是天地元素之间的人类的特定表示。

这个主题贯穿了会所的内部设计及照明设计，并影响了整个建筑的建筑结构。同时八角形的形状还易于设计师创建一种竞技场的效果，因为建筑是一个完整的圆形。

在最大限度地优化各方观众的视角、营造最佳音效的同时，OCTAGON 内还设计了许多的直角，以便更好地容纳包括 Funktion-One 环绕立体音响在内的高科技设备。在 URBANTAINER 的世界中，设计被理解成一种特定场景中的创造性交流过程，而不是因其功效而被应用的模板。一直保持这种心态做设计是一种挑战，但结果会证明加倍的努力是值得的。这一点在极具活力的亚洲大城市的建成项目中体现得尤为突出。

URBANTAINER Co., Ltd

Hayes Slade, James Slade
Slade Architecture

The design of a club space, and especially a TOP CLUB, requires consideration of the emotional nature of leisure and recreation. Success for these venues is driven by a collection of intertwined attributes that are highly subjective and often judged in an instant.

In this world, there is tremendous latitude and at the same time, there are few rules that point definitively to success. Often the appeal stems from the creative balance of seemingly contradictory qualities: welcoming/selective, cutting-edge/familiar, authentic/trendy, interesting/background. The merging of design, operations, program and service to create a unified enticing atmosphere is a complex interaction that defies codification. As a result, these clubs and hospitality spaces embody the multifaceted nature of interior design in a particularly challenging way that spans from science to logistics to art.

Part fantasy, part functionality, the club space offers a stage set for a non-scripted experience. Design at every level is central to setting the tone: from the largest scale, where flow and spatial configuration guide social interactions down to the smallest detail level where detail and material cues place visitors in a specific mindset.

会所，尤其是顶级会所的空间设计，需要考虑到其休闲娱乐的情感特性。这些场所的成功取决于一系列错综复杂的特性，而且这些特性是非常主观的，常常一瞬间的判断就决定其成败。

在会所的世界里，设计风格范围广阔，与此同时，指明成功的规则却不多。通常，会所的魅力来源于看似矛盾的特质之间创造性的平衡：大众与个别、前沿与普通、古旧与时尚、引人注目与默默无闻。将设计、执行、程序和服务集合起来，创建出一种统一且迷人的氛围，是一个复杂的融合过程，极具挑战性。因此，这些友好的会所空间以一种特别具有挑战性的方式，从科学方面到后勤再学到艺术领域，都充分体现了室内设计的多样性本质。

梦幻而又不失功能性，会所空间提供了一个即兴表演的舞台。各级的设计对于会所的定调十分重要：广义上讲，客流量和空间配置引导互动交流；狭义上讲，细节和材料将顾客定位在一个特定的心态上。

PREFACE II 序言二

CONTENTS

目录

CLUB

会所 CLUB

URBANTAINER Co., Ltd

Club OCTAGON

Team
Jiwon Baik, Younjin Jeong, Hyungsuk Lee, Semi Kim, Gay-oung Lee, Yoonyoung Chang

Location
New Holltop Hotel, Seoul, Korea

Area
2,600 m^2

4D Media Light
URBANTAINER Co., LTD+NASA Factory (James Powderly, Narim Lee, Eunjeung Son)

Photographer
Sun Namgoong

Club OCTAGON answers the client brief of renovating 2,640 m^2 over two levels of gutted hotel basement to create a high tech auditorium, club, lounge and restaurant that put music and people's experience first. URBANTAINER developed a new type of multi-space for entertainment, socializing, and subculture that was lacking in the Korean market. Conceptually every detail of Club OCTAGON works with the octagonal form from the corporate identity, including the layout, 4D Media Lighting, modular seating, to even the ice buckets in each of the VIP rooms.

The space facilities include the main floor dance floor and entertainment hall, three bars, open kitchen, mezzanine private VIP bunkers, 2nd floor VIP lounge, 2nd floor 2nd stage lounge, silent room, and women's powder room. Some of the most famous clubs in Europe utilize the raw and empty space of factories, power stations, and prisons. In these spaces, the club is a place where culture can reinvent itself and celebrate the party scenes inhabitating them.

The design concept for Club OCTAGON was based on the energy of these spaces, instead of aiming to be Seoul's next trendy club. The lack of converted warehouses and factories as creative multiuse spaces in Korea inspired URBANTAINER to create their own whilst including the glossy and technological elements that are a must for the wired Korean market. URBANTAINER considers this design storytelling, imaginatively fulfilling desires, in this case inspired by European club culture and its converted spaces, by orchestrating the mood through design. With the concept of a factory in mind, the club relies on limited materials and colors only. The minimal design with elements like exposed steel beams, elevator shaft, fire prevention system, ventilation, and epoxy cement floors to allow focus on the programmatic content of the events, while keeping alive the excitement of the raw energy of old school raves.

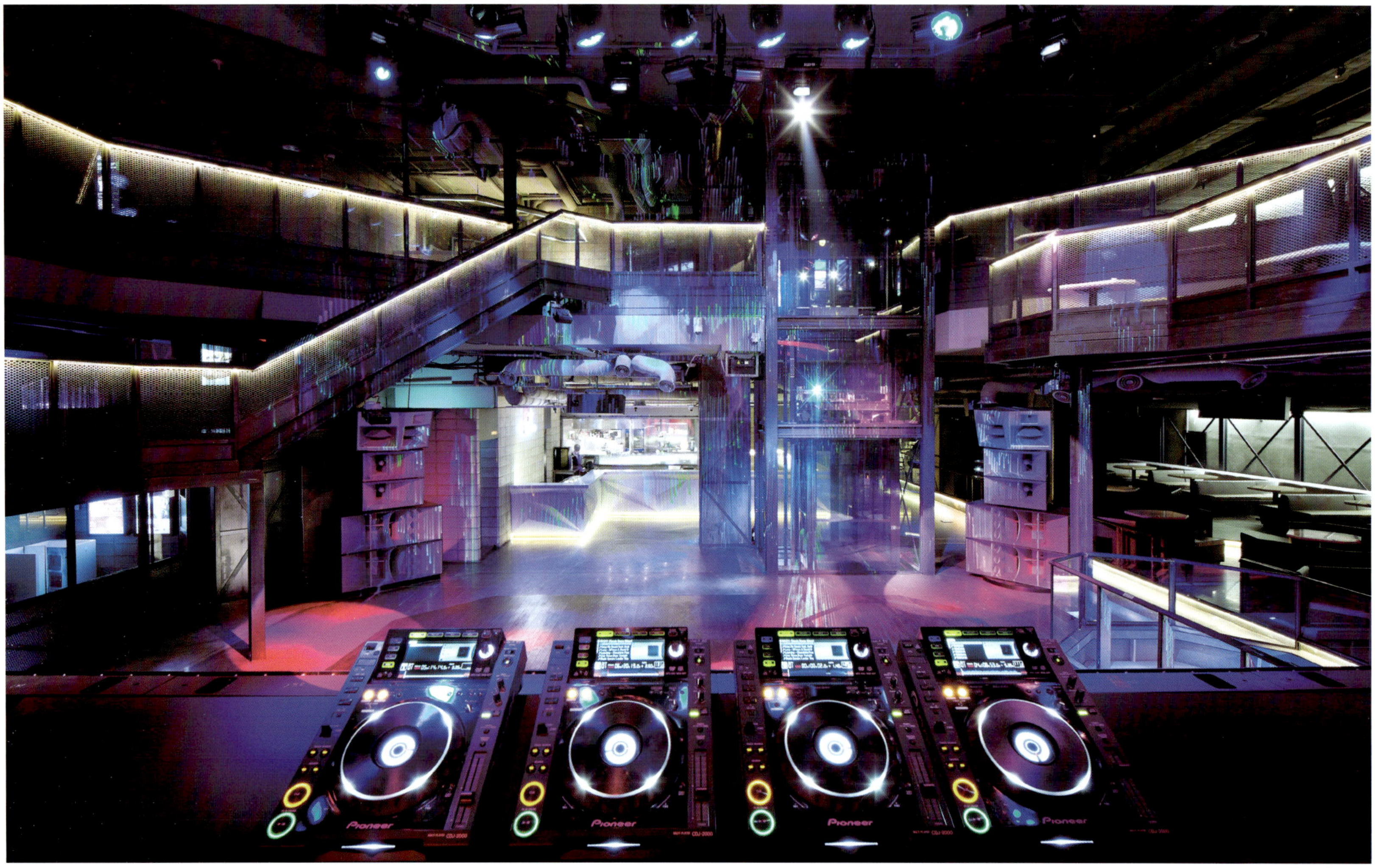

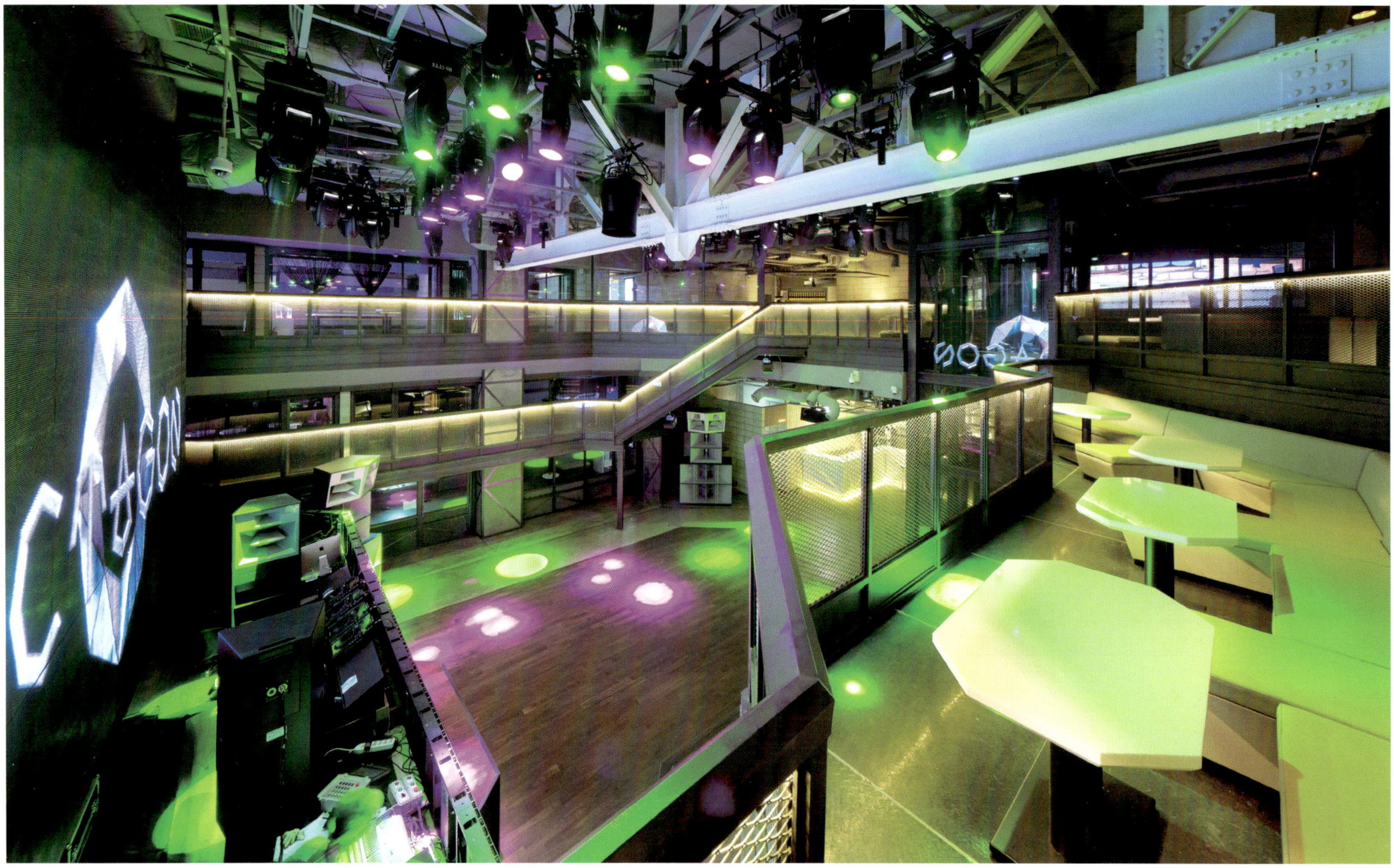

应客户要求，OCTAGON 会所翻新了面积达 2640 平方米的两层废弃的酒店地下室空间，将其改造成一个高科技礼堂、会所、酒吧和餐厅，并把音乐和人们的体验放在首位。URBANTAINER 开发出一种新型的多用途空间，用于娱乐、社交和亚文化圈，这在韩国市场上是很少见的。从概念上讲，OCTAGON 会所的每一个细节都与企业标识中的八角形相关，包括布局、4D 媒体照明、模块化座位，甚至连每个贵宾室内的冰桶都是八角形的。

会所内的设施包括主舞池、娱乐厅、三个吧台、开放式厨房、夹层 VIP 贵宾室、二楼的 VIP 贵宾休息室、二楼二期休息室、免打扰房间以及女士化妆间。欧洲一些最著名的会所会使用工厂、发电站、监狱等地的原始空间。在这些空间内，会所是一个可以文化自主革新并庆祝革新成功的派对场所。

OCTAGON 会所的设计理念基于空间的能量，而非旨在成为首尔的下一个时尚会所。在韩国，以改建的仓库和工厂作为创造性的多用途空间的实例很少，这一点启发了 URBANTAINER 创造属于他们自己的空间，该空间需要有光泽性和技术性，这对韩国的有线市场来说是一个必要条件。URBANTAINER 认为这一设计很梦幻，能够满足人们的愿望。本案中，受到欧洲会所文化及其改造空间的启发，设计通过精心策划将情绪贯穿于整个设计中。时刻牢记着工厂的概念，会所只使用了有限的材料和颜色。暴露的钢梁、电梯井、消防系统、通风装置和环氧水泥地板上寥寥几笔的设计，可让人们的注意力集中到会所内的活动上，同时还能保有守旧派大加称赞的原始力量。

ONG&ONG Pte.Ltd

The Second Floor, American Club

Team Director
Lynn Ng, Teo Boon Kiat

Location
Singapore

Photographer
See Chee Keong

The Second Floor, a new restaurant in the American Club, was designed with the intention of fusing Eastern and Western influences. It features tiled ceilings, a brick wall behind the bar, and a hardwood-floored pavilion, evoking an ambiance that is reminiscent of Singapore's colonial past. Eastern influences come into play through the use of movable dividers, bringing to mind traditional Chinese screens. By clever manipulation, these screens can be used to expand the bar area. Shades of sapphire, emerald, and amethyst dominate the interior, casting a muted and sophisticated atmosphere upon the restaurant. These neutral shades, combined with glass dividers that reach from the floor to the ceiling, exude a taste of refinement that characters of The Second Floor.

EXIT

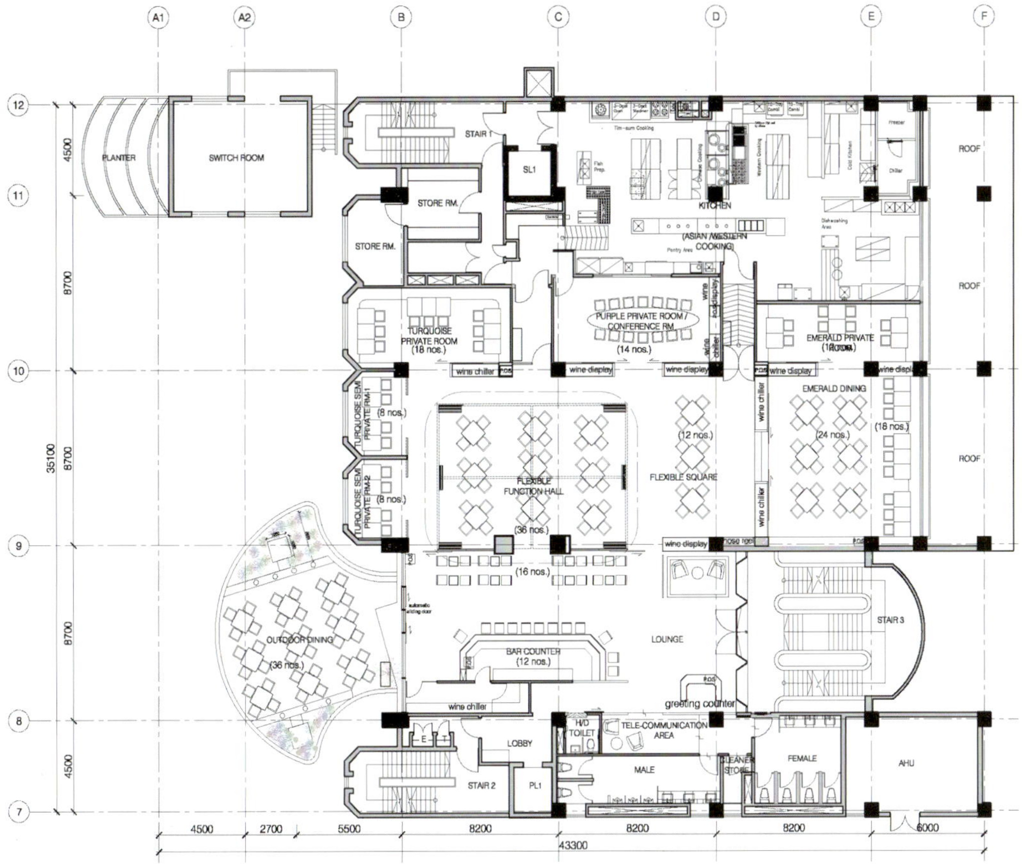

1 FURNITURE LAYOUT PLAN
FP-01 SCALE 1 : 200

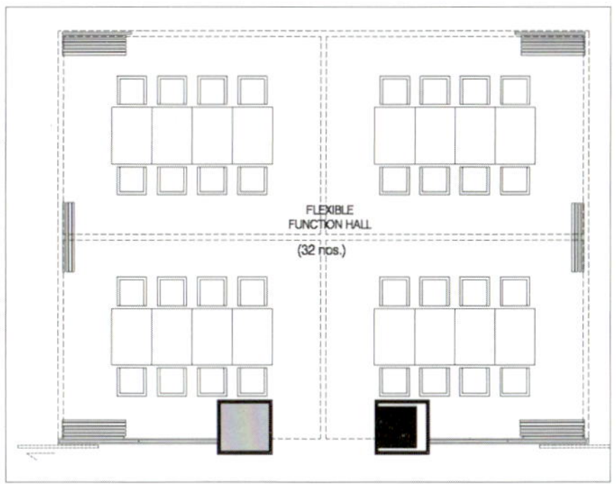

2 FLEXIBLE FUNCTION RM PLAN(OPTION -2)
FP-01 SCALE 1 : 100

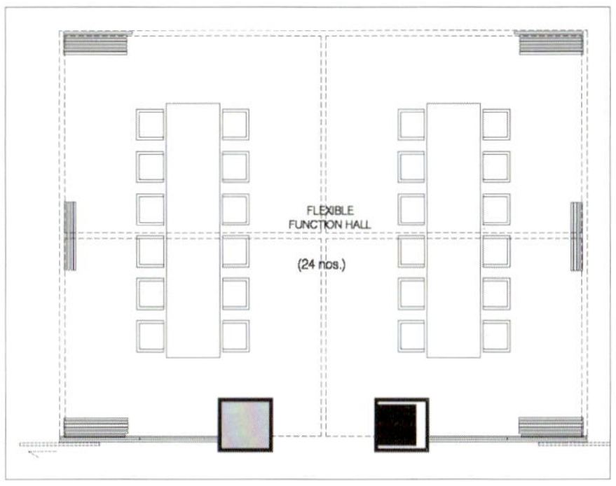

3 FLEXIBLE FUNCTION RM PLAN(OPTION -3)
FP-01 SCALE 1 : 100

The Second Floor 是美国会所的一个新餐馆，其设计旨在融合东西方的风格。瓦片天花板、吧台后的砖墙面，以及一个铺有硬木地板的凉亭是该餐馆的设计特色，餐馆的氛围会让人联想起新加坡的殖民时期。可移动的隔墙设计，会让人联想到传统的中国屏风，进而成功地渲染了餐馆的东方氛围。经过巧妙的处理，这些屏风还可以扩展酒吧区的空间。蓝宝石、绿宝石、紫水晶的渐变色调主宰着空间，为餐厅营造出一种低调且精致的气氛。这些中性色调与落地玻璃隔断相结合，散发出一种精致的味道，成为 The Second Floor 的特色。

Slade Architecture

Virgin Atlantic JFK Clubhouse

Architect
Hayes Slade, James Slade

Team
Tian Gao, David Iseri, Alessandro Perinelli, Magda Stoenescu, Frances Calosso, Yuko Okuma, Garrett Pruter

Location
Queens, New York, USA

Area
1,000 m^2

Photographer
Anton Stark, Slade ArchitectureTG

The Clubhouse is a hybrid-private members club, boutique hotel lobby, restaurant and chic bar. Bounded on two sides by expansive views over the jet ways, aircraft and the iconic TWA terminal, the lounge picks up on the glamour of the 1960s air travel with an uptown vibe. Distinct areas cater to different passenger activities and interactions. The areas are organized by acoustical zones: Quiet lounge, Talking lounge, cocktail lounge. Activities that require less time are close to the entry, those that require more are further.

Cloud shaped in plan, the central cocktail lounge is enclosed by a diaphanous, curving screen of stainless steel rods and walnut fins. The rods and fins mediate views and create an internal skyline. 2,000 powder-coated cylinders hang from the ceiling creating a glowing, sculptural topography. Guests move through the cocktail lounge in a rhythmic syncopated flow to the many unique Clubhouse amenities and spaces defined by this central element.

Furnishings help define the character of each area. We designed the grey amorphous pebble sofas and a red ball sofa to create a seating landscape in the cocktail lounge. The talking lounge includes the restaurant and is furnished to encourage groups and interaction. The quiet lounge emphasizes individual seating. Aluminum walls perforated with a pixelated cloud pattern are cutout to create "floating" seating pods. Spa and hair salons complete the amenities. Details reinforce the "Uptown" theme of the lounge. We designed custom wallpaper composed of New York icons: the Chrysler Building and the Empire State. In the bathrooms, white subway tiles and large-scaled black and white Sanborn maps of NYC complete the picture.

这是一家功能多样的私人会员会所，拥有精品酒店大堂、餐厅和别致的酒吧。会所两侧景观开阔，可以观赏到喷气式飞机滑道、飞机和标志性的环球航空公司候机楼。会所重现了 20 世纪 60 年代航空旅行与住宅氛围相结合的独有魅力。不同的区域迎合着不同旅客活动和社交的需求。各区域依据声学标准划分：安静的休息区、交谈区和鸡尾酒区。耗时短的活动靠近入口，需要较长时间的活动则在里面。

会所的平面图呈云形，中央鸡尾酒区四周是透明、弯曲的屏风，屏风由不锈钢管和胡桃木片组成。不锈钢管和胡桃木片能够调节视线，创造一个内部的天际线。2000 个粉饰圆筒悬挂在天花板上，营造出一种热情洋溢、雕刻般的形象。在这一中心装饰元素下，客人有序地穿梭在鸡尾酒区，享受更多独特的会所设施。

家具定义了每个区域的个性。在鸡尾酒区，设计师设计了灰色的形状各异的“卵石”沙发和红球沙发，二者组成了主座位区。交谈区配置了餐厅，便于鼓励群体活动和互动。安静的休息室则以单人座椅为主。铝制墙体穿孔，孔洞呈云形，以创建“浮动”的座位。水疗和美发沙龙的设置进一步完善了会所的功能。细节设计强化了休息室的“住宅”主题。设计师为会所量身定制了由纽约标志性建筑（克莱斯勒大厦和帝国大厦）组成的壁纸。在浴室内，白色的铺地瓷砖和大幅黑白的纽约桑伯恩地图更加完善了“住宅”的主题。

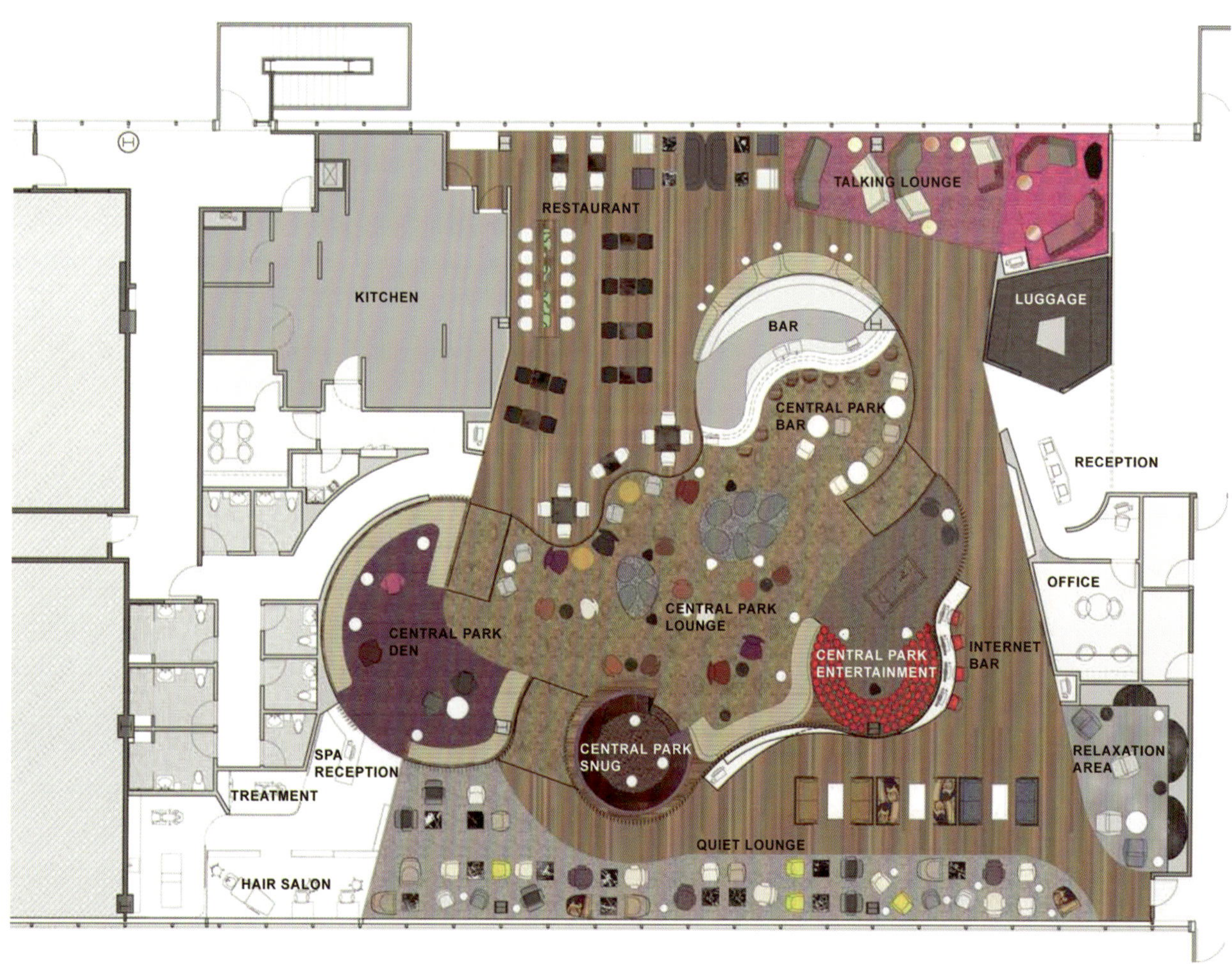

TALKING LOUNGE
RESTAURANT
KITCHEN
BAR
LUGGAGE
CENTRAL PARK BAR
RECEPTION
OFFICE
CENTRAL PARK DEN
CENTRAL PARK LOUNGE
CENTRAL PARK ENTERTAINMENT
INTERNET BAR
SPA RECEPTION
CENTRAL PARK SNUG
RELAXATION AREA
TREATMENT
QUIET LOUNGE
HAIR SALON

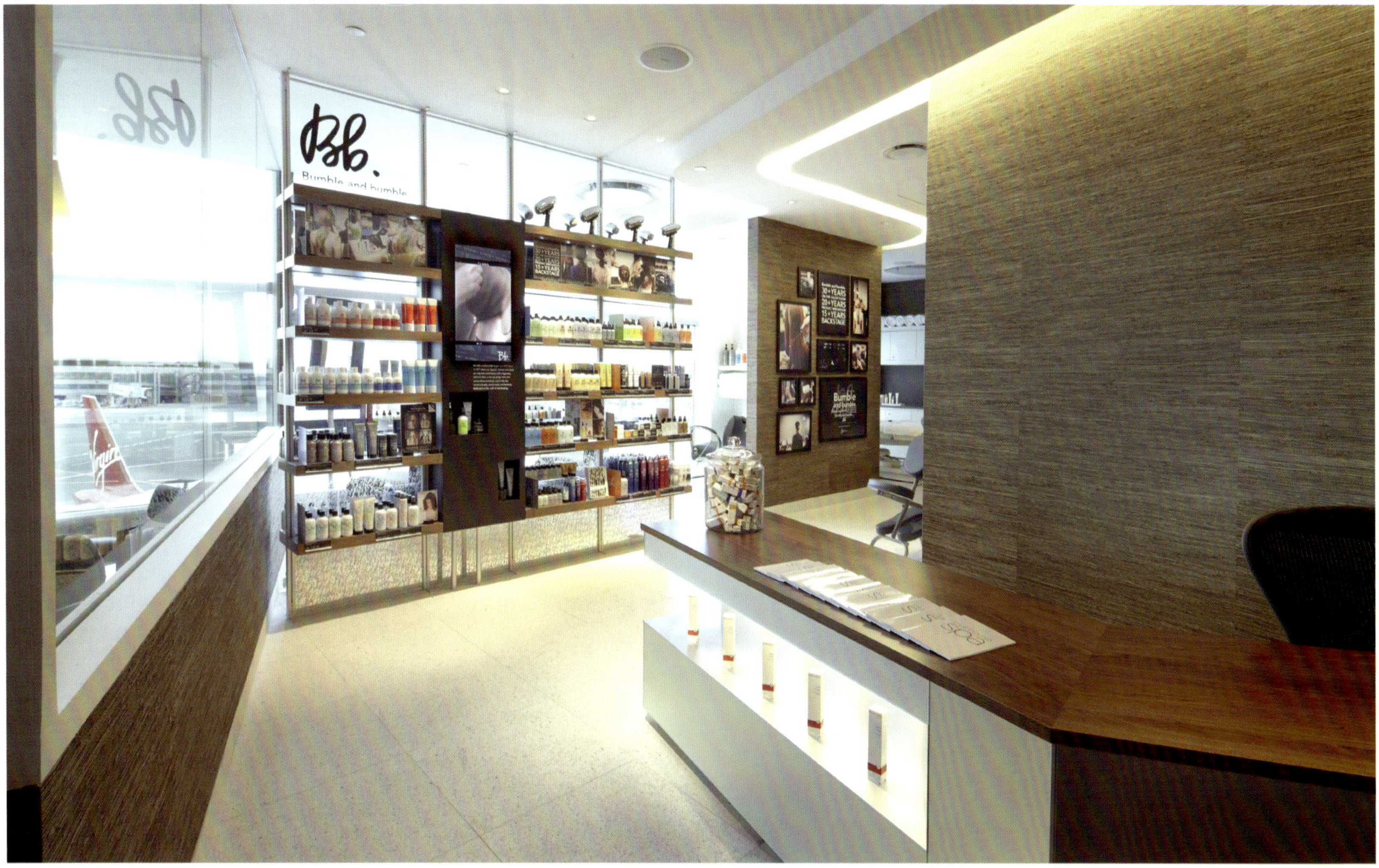
Bb.
Bumble and bumble

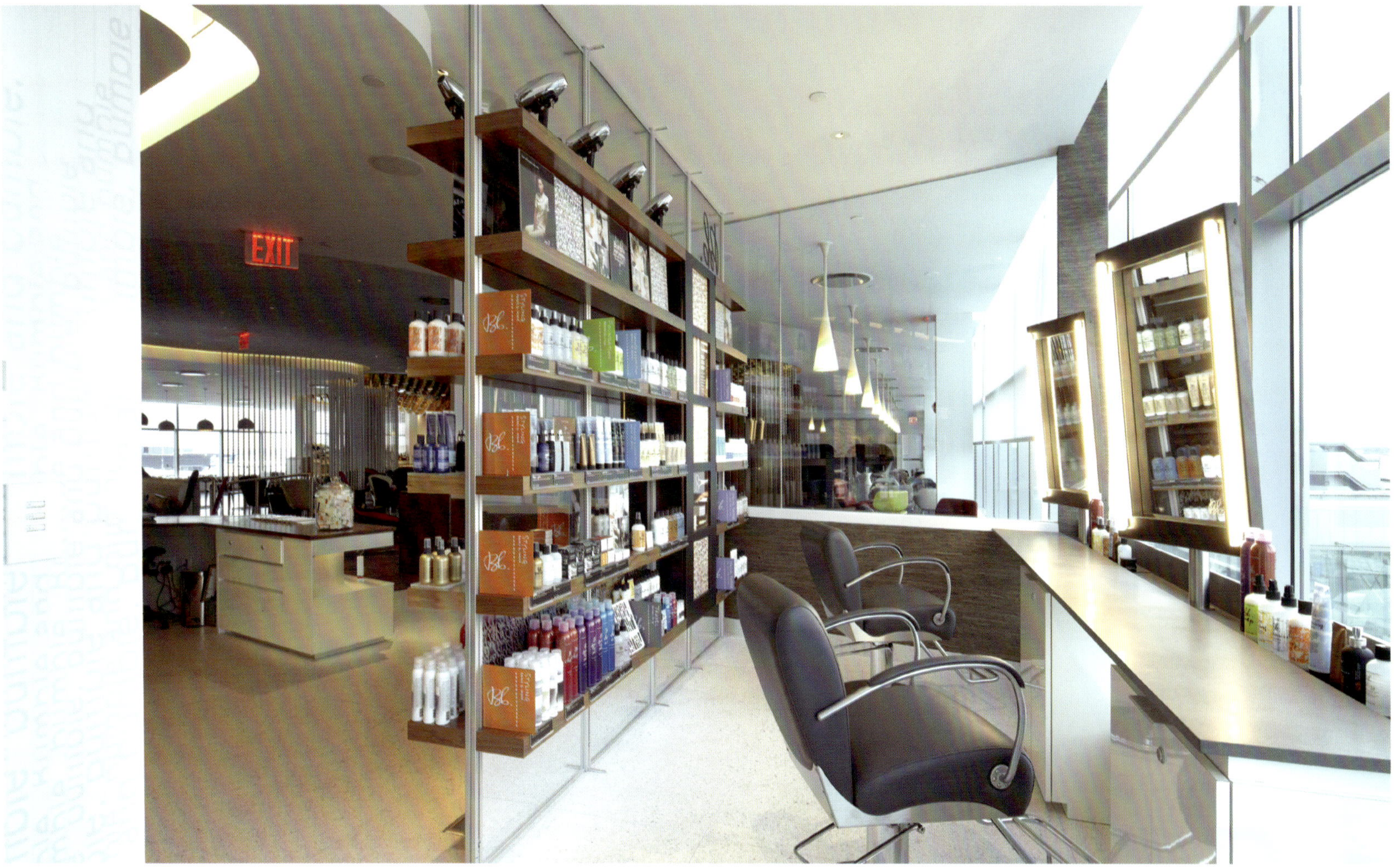
EXIT

COLUMBUS

dwp | design worldwide partnership

Chill Skybar & Restaurant

Location
Ho Chi Minh City, Vietnam

Area
944 m^2

Designed by world-class architecture and interior design firm dwp, Chill is a stunning new rooftop bar and restaurant – a stylish addition to Ho Chi Minh City. Juxtaposed from the streamlined, robust character of the office building, upon which it perches, this venue enhances the enjoyment of all sensual pleasures with its curvaceous, fluid, urban drama.

Chill crowns AB Tower, comprising two levels – an upper lounge and lower dining area. Smoked mirror walls create an illusion of endless space in the small lobby, where guests are greeted and invited indoors or out. The outdoor cigar lounge is where live jazz plays over resplendent views and warmly-toned color changing lights. The indoor lounge area overlooks both the city and the dining area below. The VIP room, located at the end of the lounge, offers the best view spot, with a private balcony and shimmering drapery.

The curved facade and double-height, slanted wall glazing echo intrigue and impact, with a series of vertical steel structures, spanning the length. Levels are linked by a grand staircase, penetrating through the glazing wall out to the dining area. The self-illuminating spiral sky bar keeps guests entertained, with the performance of expert mixologists, and resident DJ close by. The aqua-blue tinted lighting of the curved glass balustrade generates an instant landmark from street level.

Chill 由世界一流的建筑与室内设计公司 dwp 设计，是一家绝妙的新式屋顶酒吧和餐厅——为胡志明市增添了一道时尚靓丽的风景线。Chill 位于办公楼之上，与流线型、粗犷的办公楼相呼应，其曲线形的体量、流畅的设计和城市化的风格可使顾客在此体会到极佳的感官享受。

Chill 以 A、B 两塔为标志，由上层会所和下层餐厅两部分构成。小厅中的烟熏镜墙营造了一种错觉，使空间看起来无限延伸，客人可从此处进入餐厅或离开。在户外的雪茄吧里，伴随着华美的景观和暖色调变化的灯光，正上演着现场爵士乐表演。室内休息区可俯瞰城市和底层的餐厅。VIP 房间坐落于会所的一端，配有私人阳台及闪闪发光的的帷帐，在这里能够欣赏到最好的景色。

弧形的外立面、双层高的斜玻璃墙与一系列垂直的钢结构遥相呼应，为餐厅营造了一种神秘与冲突的效果。楼层由豪华的楼梯连接在一起，楼梯穿过了玻璃墙一直延伸到餐厅区域。自身可发光的螺旋空中酒吧、专业调酒师的表演以及附近的驻唱 DJ，这些都令客人感到愉悦。从街道上看去，弧形玻璃扶栏的水蓝色照明灯是一道显眼的标志。

Chill

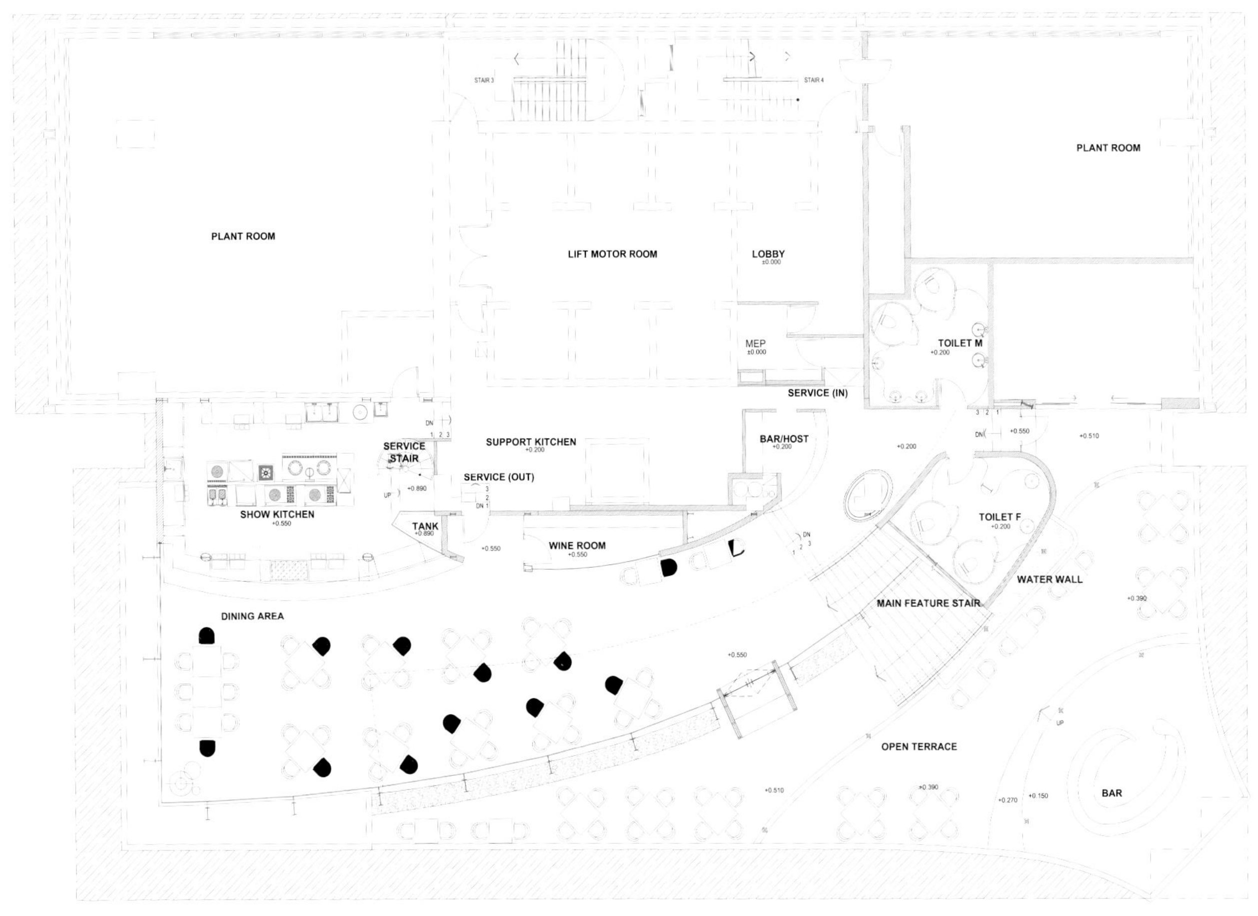
PLANT ROOM
LIFT MOTOR ROOM
LOBBY
PLANT ROOM
MEP
TOILET M
SERVICE (IN)
SERVICE STAIR
SUPPORT KITCHEN
BAR/HOST
SERVICE (OUT)
SHOW KITCHEN
TANK
WINE ROOM
TOILET F
WATER WALL
MAIN FEATURE STAIR
DINING AREA
OPEN TERRACE
BAR

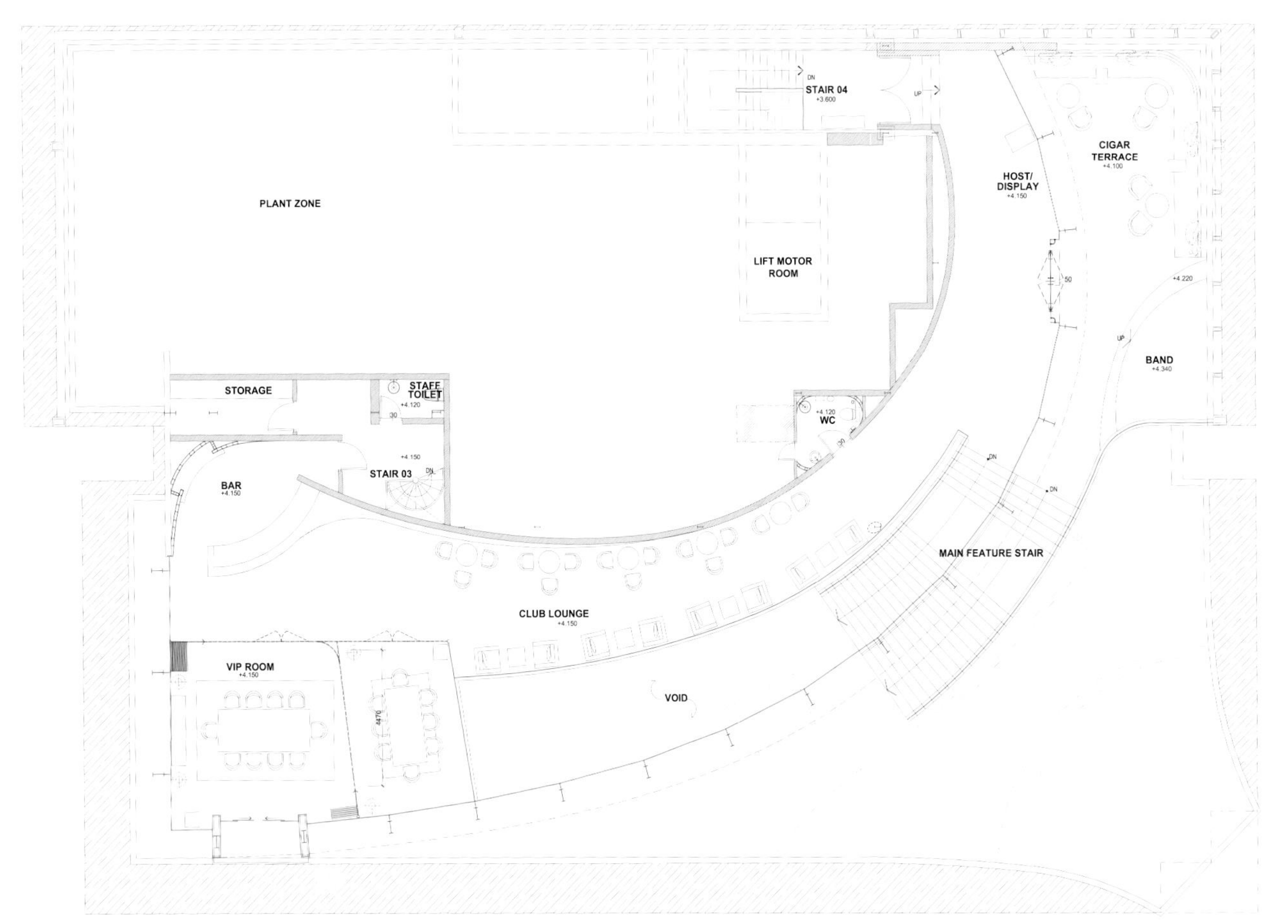
STAIR 04
CIGAR TERRACE
HOST/ DISPLAY
PLANT ZONE
LIFT MOTOR ROOM
BAND
STORAGE
STAFF TOILET
WC
BAR
STAIR 03
MAIN FEATURE STAIR
CLUB LOUNGE
VIP ROOM
VOID

Mr. Important Design

The mix of vintage and new, the desire to push the use of materials, the need to transform the interior from restaurant to club ambience and the evocation of a kind of Asian feeling of dark glamour, these were the objectives in the design of Motif.

Designer mixed things up by using futuristic materials such as stretched glossy polymer film to provide an undulating ceiling of reflected color and pattern in the DJ area. We used cutting edge Asian furnishings with traditional materials from Kenneth Cobonpue coupled with the modernism of Konstantin Grcic all mixed up with the 1960's and 1970's glam of Tommi Parzinger style pendants and sleek black linear silk shades.

Hundreds of feet of black and silver chain hand worked into a bold floral tapestry (art provided courtesy of Amy Butler) that envelopes the space without suffocating it. Yards and yards of thick black silks rope to soften and provide pattern to the institutional cinder block construction of the buildings interior.

Location
San Jose, CA, USA

Area
743 m^2

Extensive and saturated use of color changing LED lighting to transform the intentionally monochrome palette of Motifs' interior into a heady polychromatic club experience later in the evening. One of the main challenges was to make the switch from restaurant to club with just the ambient lighting. We created an entirely monochrome interior that would be appropriate for daytime and evening function. Early on the restaurants monochrome furnishings were bathed in warm amber glow from Edison bulbs in black silk shades. Later in the evening the full effect of the extensive LED system would flood the monochrome interior with polychromatic lighting and slow changing effects would herald the beginning of serious nightlife.

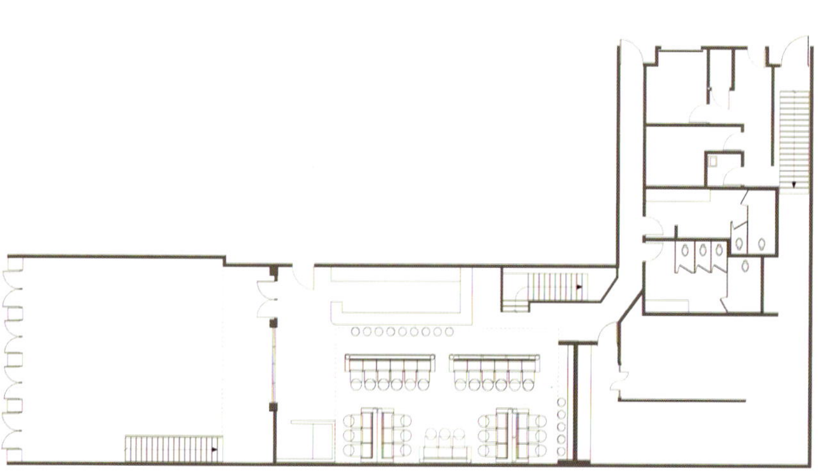

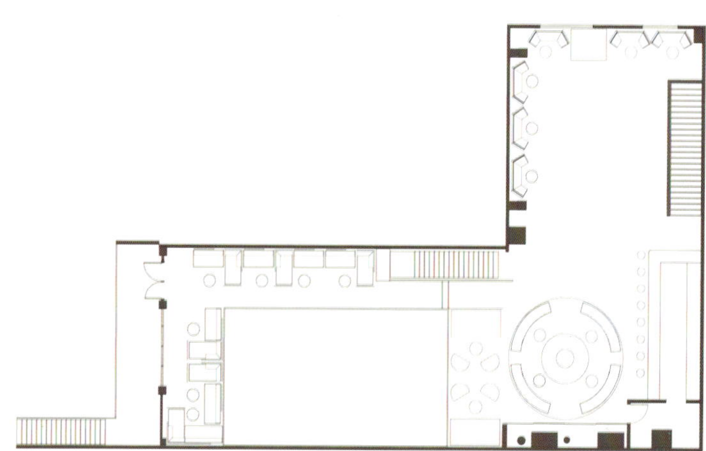

Motif 的设计宗旨是：融合古典元素与时尚元素，充分利用装饰材料，将餐馆氛围改造成室内会所氛围，以及重现一种亚洲的黑色魅力。

Motif 的设计师善于运用未来主义材料并将其混合运用，例如运用延伸的光滑的高分子薄膜，在 DJ 区营造一个能够反射各种颜色和图案的波浪状天花板。设计师运用了最尖端的亚洲陈设，加上 Kenneth Cobonpue 的传统材料和 Konstantin Grcic 的现代主义风格，所有的这一切都与 20 世纪 60 年代到 70 年代的 Tommi Parzinger 风格的华丽垂饰和光滑的黑丝绸线形灯罩完美地融合在一起。

手工制作的数百英尺长的黑色和银色的链子组成了一个宽大的花型织锦（感谢 Amy Butler 的艺术提供），包围了整个空间，却没有令人窒息的感觉。一段段的黑色粗丝绳被用来软化空间，并为会所室内的煤渣砌块结构提供了图案。

夜晚时分，大量使用色彩饱和度高的可变色的 LED 灯，以转换室内氛围，使 Motif 原本单色的室内装潢变成了令人兴奋的多彩的酒吧氛围。仅仅利用环境灯光照明将餐厅氛围改造成会所氛围，这是设计面临的一个主要挑战。为此设计师营造了一个完全单色的室内设计，以同时满足日间和夜间的功能需求。在一天较早的时候，黑丝绸灯罩下的爱迪生灯泡发出琥珀色的灯光，餐厅单色的陈设沐浴在温暖的黄色光辉中。当夜幕降临，在强大的 LED 灯光系统下，多彩的灯光将这些单色陈设淹没起来，缓慢变换的灯光效果也在宣告着夜生活的开始。

Slade Architecture

Virgin Atlantic EWR Clubhouse

Architect
Hayes Slade, James Slade

Team
Tian Gao, Kristina Kesler, Magda Stoenescu, Rasmus Kristensen, Ana Lopis, Allesandro Perinelli, Garrett Pruter

Location
Newark, New Jersey, USA

Area
500 m^2

Photographer
Anton Stark

The Virgin Atlantic EWR Clubhouse brings together NYC's downtown flair and Virgin Atlantic's warmth and individuality. A constellation of sparkling pinpoint lights draws you into the sculpted entry passage. Punctuated by a sky lit receptionist at a specular desk, the entry sequence creates an extended decompression zone between the busy terminal and the relaxed luxury of the clubhouse; a dramatic immersion into the Virgin Atlantic experience.

The Clubhouse is organized around a faceted bar in the center of the lounge, culminating in a crystalline bottle display under the central skylight. The geometry of this object ripples across the ceiling; an abstract geometric canopy that spills up into the two skylights. A series of distinct spaces pin wheel around the bar. Each space references an iconic downtown typology – café, theater, art gallery, restaurant, club, bar, lounge. Grab a coffee and read a book or magazine in the wood lined café. Venture into the curtain wrapped screening room to catch the latest art house video. Curl up in an upholstered pod, carved into protective faceted concrete of the passion pit. Pull up to the bar at the liquid lounge. Sit down for a delicious meal in the gallery-like brasserie. Catch up on your email or listen to music in the colorful origami lounge.

Art is central to the downtown vibe. Digital finger paintings of lower Manhattan by New York artist Jorge Colombo, animate the café. The brasserie features three mixed media works by New York artist Garrett Pruter, inspired by discarded travel photographs found in various downtown junk shops. Works by Kate Hazell, Galia Rybitskaya, Mille Marotta and Jonny Moss are displayed in the space. Custom elements abound, adding moments of discovery and humor. Pillows examine the urban landscape at different scales. Graffiti inspired wallpaper in the bathroom/shower area is illuminated by glowing resin sinks. The more time you spend exploring the lounge, the more it unfolds and reveals itself.

The Virgin Atlantic EWR 会所汇集了纽约市中心的魅力以及维珍大西洋航空公司的温馨和个性。璀璨闪耀的定位灯光将你带进雕刻般的入口通道。接待员站在反光柜台后面，上方由天窗提供自然照明，你会发现入口通道在忙碌的候机楼和休闲豪华的俱乐部之间扩展出一个减压区，使得游客戏剧般地沉浸到维珍大西洋航空公司的服务中去。

会所围绕着空间中心的一个棱角分明的吧台设计，吧台不断向上，最终在中央天窗下形成一个透明的酒品展示台。展示台的几何状轮廓倒映在天花板上，抽象的几何形倒影，看上去像是两个天窗。一系列不同的空间围绕着酒吧而设。每个空间的设计都参照了一个标志性的市中心建筑类型——咖啡馆、剧场、美术馆、餐馆、会所、酒吧、休息室。你可以在木制的咖啡馆内喝一杯咖啡，读一本书或杂志；可以进入由窗帘围合而成的摄影室观看最新的艺术剧院视频；也可以蜷缩在铺着软垫、嵌在混凝土中的几何状小窝内沉思。在酒吧小酌一杯，坐在类似画廊的啤酒店内享受美味佳肴，在五颜六色的折纸休息室内查看邮件或听音乐。

艺术是会所的核心氛围。由纽约艺术家 Jorge Colombo 创作的下曼哈顿区的数码触摸画，使整个咖啡馆充满生机。啤酒店的特色是纽约艺术家Garrett Pruter创作的三个混合多媒体作品，作品的灵感来自于在市中心各种旧货商店里发现的废弃的旅行照片。Kate Hazell，Galia Rybitskaya，Mille Marotta 以及 Jonny Moss 的作品都陈列在这个空间内。自定义元素比比皆是，引人探索、妙趣横生。靠枕在不同程度上体现着城市氛围。浴室 / 淋浴区涂鸦式的壁纸被发光的树脂水槽照亮了。越多地探索这间俱乐部，它就会呈现越多。

Virgile and Stone / Imagination

Morton's Club

Location
London, UK

The club has a rich history dating back decades and the aim of the extensive refurbishment was to reflect the area's new more fashionable character and attract an exclusive but younger membership base.

The design reflects the heritage of the club's location and history as its origins as a private residence of a British aristocrat. In addition to the design challenges intrinsic to the new concept for the club – that of creating a contemporary environment that will appeal to existing as well as new members – the project's team were met with a number of design challenges as the Georgian building is a Grade II listed building.

The design scheme respectfully enhanced the existing Georgian features and re-instated those previously neglected. Vibrant British eclecticism seamlessly combines elegant traditional features offset with a modern twist, creating a feel of a "lived-in", effortless and relaxed grand interior, housing a carefully curate collection of art.

Timber paneling has been re-introduced to the bar, hand finished in deep Auvergne color, complementing the impressive new bar, which is fabricated from bespoke hammered brass. Custom designed cocktail cabinets stand behind the bar and when opened, show an edge lit modern interior. The grand proportions of the restaurant and views onto Berkeley Square are framed in a distinctive grey green.

Bespoke designed furniture pieces, such as banquettes, credenzas and specially made light fittings add to the desired residential feel. Walls display an extensive and carefully curated collection of original contemporary artworks by British artists, from Howard Hodgkin to Julian Opie, discreetly lit and creating an intimate environment. Down in the Lower ground floor the atmosphere becomes more intimate as entering a customized club-style area with exposed brickwork, a DJ booth and jet-black bespoke tables that hold ice chillers. The night lounge raw brick lined walls are encased within glass panels that reflect and refract the color changing light and moving digital artwork.

Morton 会所已有几十年的光辉历史，这次大规模翻新的目的是为了反映该地区更时尚的新特点并且为吸引更年轻的专属会员打基础。

设计反映出了该会所的位置和历史的起源，这里曾是一座英国贵族的私人住宅。设计除了从本质上挑战会所的新概念——营造一种现代环境，以迎合现有的及未来的会员——项目团队还遇到了一系列其他的设计挑战，譬如乔治亚建筑是国家二级保护建筑。

设计方案充分尊重并强化了现有的乔治亚建筑的建筑风格，并重新定义了之前被忽视的建筑特色。充满活力的英国折衷主义手法将优雅的传统特色与现代手法天衣无缝地结合起来，营造出一种“天生一体”的感觉，宽敞舒适的室内空间中摆放着众多珍藏艺术品。

木镶板被重新引入到酒吧设计中，手工将其涂刷成深色的奥弗涅风格，使这家由定制的锤炼黄铜制作而成的新酒吧更具魅力。定制的鸡尾酒橱柜位于酒吧后面，当酒柜打开时，人们会看到边缘照明的现代化内部设计。餐馆的大部分景观和伯克利广场的景观都是与众不同的灰绿色格调。

定制的家具配件，如沙发、餐具橱和特制的灯具等，都为空间增添了住宅的色彩。墙面上展示了大量精美的画作，都是英国当代艺术家（从霍华德·霍奇金到朱利安·奥佩等）的原创作品，每幅画都配有相应的灯光，营造出一种温馨的氛围。进入地下室层，就进入了独有的会所风格区域，此处氛围更加温馨，顾客能够看到裸露的砖块、DJ 台和放有冰块冷却器的定制黑色吧台。这间夜总会的砖坯墙外包裹着玻璃嵌板，可以反射和折射不断变换的灯光和移动的数字艺术作品。

I may be right, I may be wrong
But I'm perfectly willing t
That when you turn'd and smiled at
A Nightingale sang in Berkeley

Pioneer

Dariel Studio

Kartel

Designer Director
Thomas Dariel

Location
Shanghai, China

Area
183 m^2(fourth floor)
186 m^2(fifth floor)
186 m^2(terrace)

Material
ceramic tile, wood floor, glass, painting

Photographer
Derryck Menere

With a concept of "Destroy Chic", the designer Thomas Dariel wanted Kartel to play with these contrasts, to be elegant yet provoking in this very heritage district.

As the demolition of the previous space was going on, the "Destroy" part of the concept naturally grew on. While the walls were falling, the original structure of the building reveals itself, featuring shapes, textures, memories that ought to be kept. As the "Beaux-Arts" style in Europe pays homage to the buildings by ripping off everything that is hiding their original architecture, the creators decided here to keep several places as original. Exposed concrete walls, undressed stripped down pillars, and old letterings made by the people who built the place are part of these heritage elements. Answering and balancing this raw side, Kartel offers an elegant and sophisticated décor served by comfy and stylish custom-made furniture, a warm atmosphere combining and mix-matching European and Asian influences. The lounge spans on three floors, two indoor plus a rooftop terrace. Three spaces offer three distinct styles and atmospheres. With moldings and fireplace, the 4th Floor reminds the cozy and plush atmosphere of the traditional. The rooftop terrace with sofas and high-top table's settings is set apart by the destructive full 360-degree views over the surrounding French Concession, a nice draw for those seeking a more casual atmosphere.

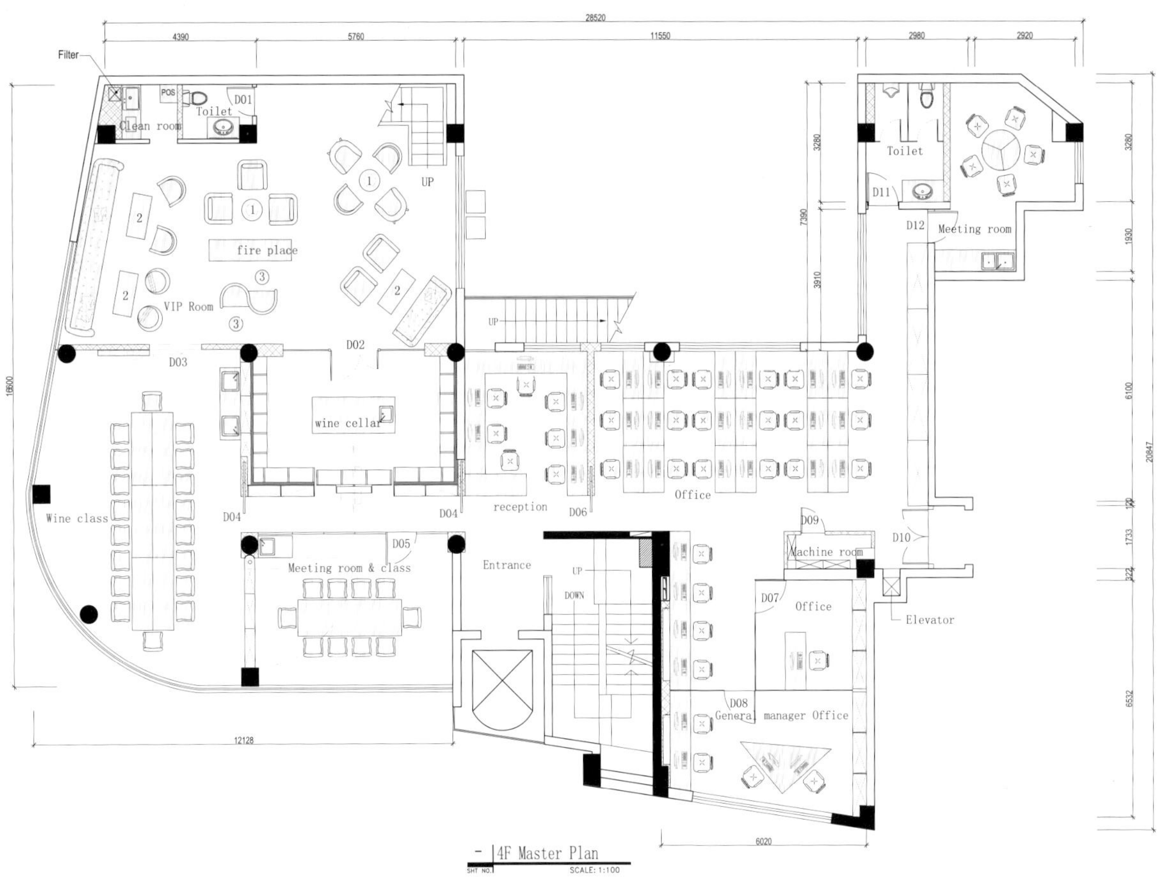

4F Master Plan

SCALE: 1:100

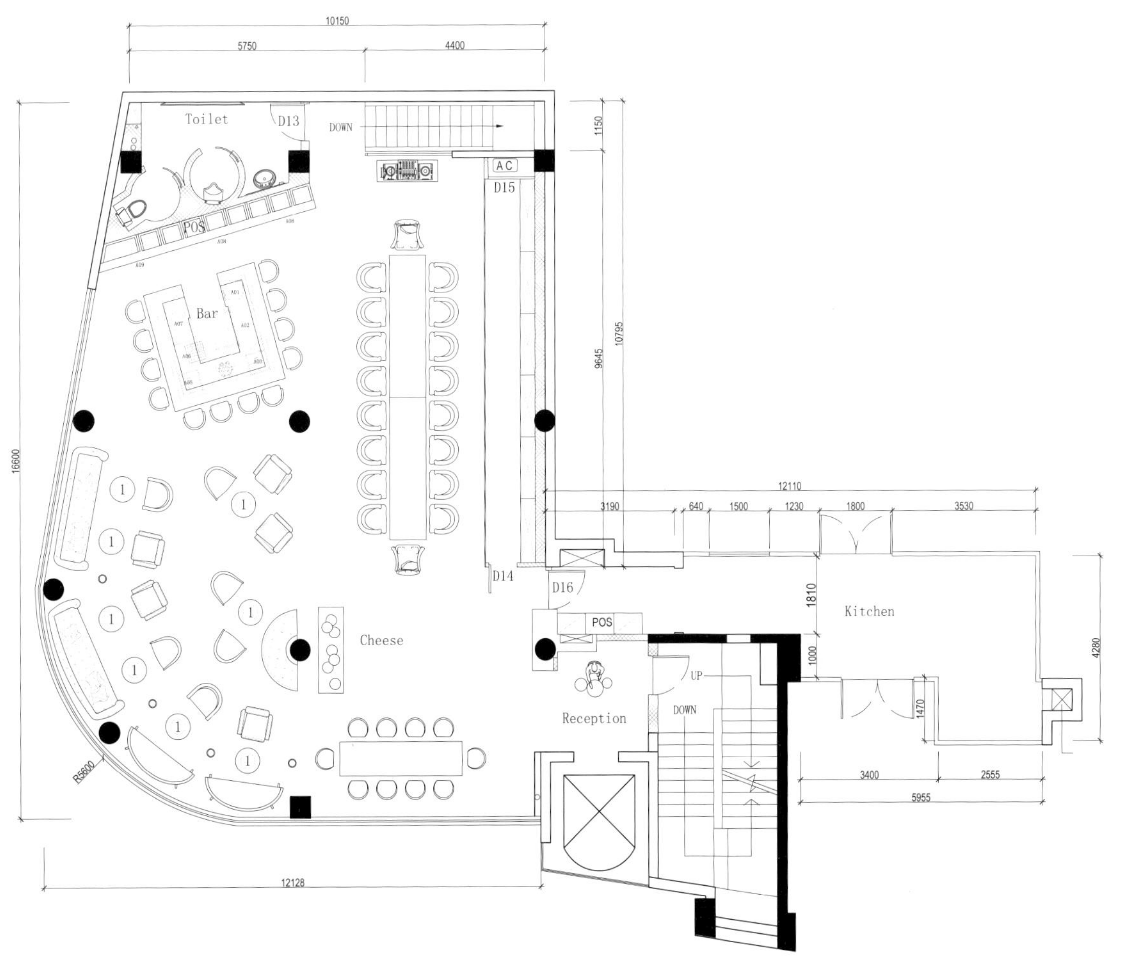

设计师 Thomas Dariel 以“颠覆时尚”为概念，希望将 Kartel 这一位于遗产保护区内的会所设计得既优雅又具有爆发力，充满对比性。

随着原空间的拆除，“颠覆”概念部分的设计也在有条不紊地进行着。随着墙面的拆除，建筑原结构慢慢显现出来，那些代表着建筑的整体形态、质感和记忆的地方都被保留了下来。欧洲“美学艺术”通过去掉建筑装饰、还原建筑原结构的方式来向建筑表示敬意，因此，设计师决定在该建筑中保留几处原结构。裸露的混凝土墙、未加工的柱子和建筑工人在柱子上刻下的文字都是这个会所传统元素的一部分。为了与这一传统部分相呼应、相平衡，Kartel 选择了一种优雅精致的装修方式，室内配有舒适且时尚的定制家具，洋溢着温馨的气氛，同时还体现出了欧洲和亚洲元素的混搭风格。休息室分布于三个楼层上，分为两个室内休息室和一个屋顶露台休息室。三个空间展示了三种不同的风格和氛围。四楼配有摆件和壁炉，让人感受传统装修所呈现的舒适豪华的氛围。屋顶露台的沙发和高脚桌位于一侧，能够 360 度观赏法租界的全景，是那些喜欢休闲氛围的客人的不错的选择。

dwp | design worldwide partnership

Capital Club, Bahrain

Location
Bahrain

Area
2,000 m^2

Photographer
courtesy of dwp

Designed by world-class architecture and interior design firm dwp, the Capital Club is Bahrain's premier private business club, offering its members exceptional meeting, networking, dining and entertaining opportunities. Set over two exquisitely luxurious penthouse floors, on levels 51 and 52 of the East Tower in Bahrain's Financial Harbour, the Capital Club, Bahrain, opened in April 2009.

The design needed to meet the demands of those from the top echelons of business, finance and government, who are powering the city's growth and turning Bahrain into the international business hub of the Middle East, that it has rapidly become known for.

dwp set the design to feature dining rooms, a bar, a lounge, a library and private meeting facilities, which all have their very own unique look and feel. dwp created and specified the finest materials, fittings, fixtures and furniture, to ensure a complete sense of luxury at this prime venue for its valued members. Rich deep colors conspired with plush, luxurious textures to generate a total immersion into class, style and sophistication, with a decidedly affluent air.

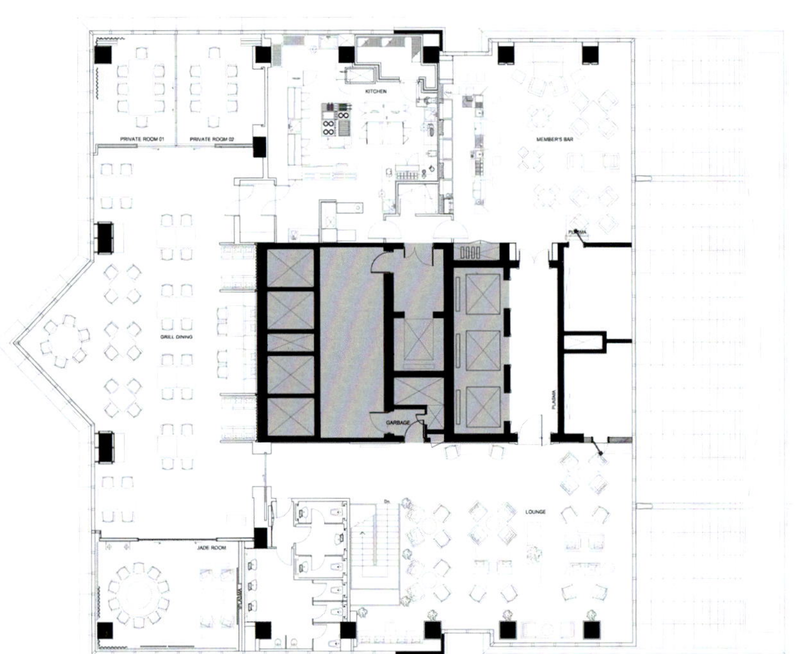
PRIVATE ROOM 01
PRIVATE ROOM 02
KITCHEN
MEMBER'S BAR
PLASMA
GRILL DINING
GARBAGE
PLASMA
LOUNGE
JADE ROOM
PLASMA

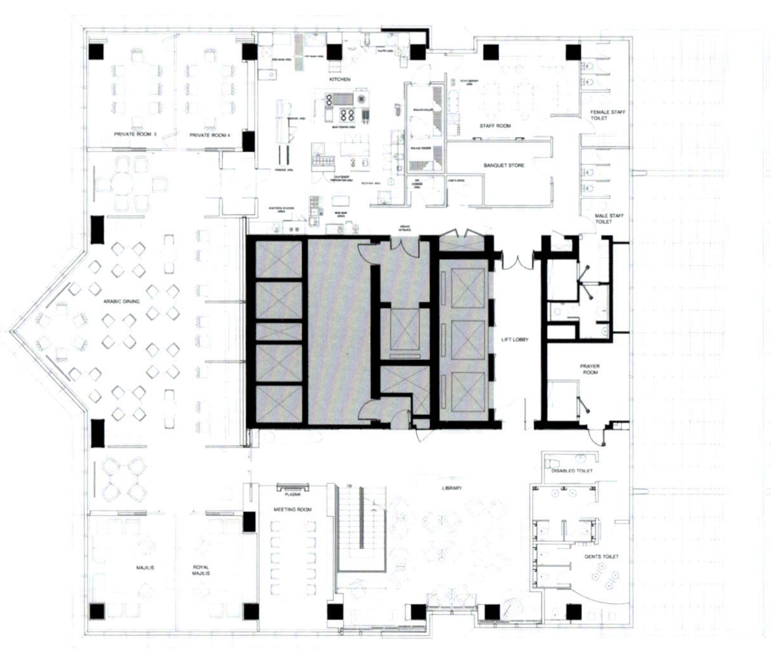
KITCHEN
PRIVATE ROOM 3
PRIVATE ROOM 4
STAFF ROOM
FEMALE STAFF TOILET
BANQUET STORE
MALE STAFF TOILET
ARABIC DINING
LIFT LOBBY
PRAYER ROOM
DISABLED TOILET
LIBRARY
PLASMA
MEETING ROOM
MAJILIS
ROYAL MAJILIS
GENTS TOILET

巴林岛 Capital 会所由世界一流的建筑和室内设计公司 dwp 设计而成，它是巴林岛的顶级私人商业会所，为其会员提供特别会议、网络、会餐和娱乐服务。巴林岛 Capital 会所于 2009 年 4 月开放，位于巴林岛金融港东塔的 51 层和 52 层（两层精美奢华的顶层公寓）。

该设计需要满足来自商业、金融和政府领域的高端要求，因为这些机构因推动了整个城市的发展，并且将巴林岛成功转型为知名的中东国际商业中心而蜚声国际。

dwp 设计了特色餐厅、酒吧、休息室、图书馆和私人会议设施，让这些空间都拥有自己独特的外观和感觉。dwp 设计并选用最好的材料、配件、饰物和家具，以确保这个顶级的会所能够为其尊贵的会员提供一种完美的奢华感觉。多种深色调与舒适、豪华的材质相搭配，无论从级别、风格还是技艺上，都使整个俱乐部沉浸在一种绝对奢华的氛围中。

dwp | design worldwide partnership

Capital Club, Dubai

Location
Dubai

Area
2,000 m^2

World-class architecture and interior design firm dwp was commissioned to redesign two F&B outlets, at the luxurious Capital Club, in Dubai. The aim was to create a pan-Asian restaurant, transforming the existing bistro space and snooker room into a very contemporary venue, and also revamp the existing Arabic dining room, Al Hamra.

dwp welcomed the challenge and designed a space that became a dynamic addition to the other F&B outlets in the Capital Club. Several limitations were encountered with respect to retaining existing ceilings and services, which were, nevertheless, overcome with finesse.

For the pan-Asian venue, the ceilings were clad with high lacquer panels to accentuate height. Backlit orange glass and natural onyx detailing emphasized the mood, while graphics created accents and added dynamic to the space. Selective walls were panelled with orange stained stingray, to provide texture. The result was a stylish, contemporary restaurant and bar, tailored to suit both relaxed lunchtime and cool evening dining. In the Al Hamra dining room, dwp designed a long juice bar, as the dominant feature in the space, glowing various tones of blue to magenta. This is highlighted by special ceiling light fixtures, also designed by dwp and manufactured in Turkey. The walls are textured stone in an Arabic motif and adorned with accompanying brass light fittings. The space is modern, with a bright, yet sophisticated, outlook.

世界一流的建筑和室内设计公司dwp受委托为豪华的迪拜Capital会所重新设计两家连锁餐厅。其目的是创建一个泛亚餐厅，将现有的小酒馆和桌球室变成一个非常现代化的场所，同时翻修现有的阿拉伯餐厅阿尔哈姆拉。

dwp勇于接受挑战，并且其设计的空间相对于其他迪拜Capital会所的连锁餐厅来说，是一道亮丽的风景线。设计在保留现有天花及服务方式方面，遇到了一些限制，但最终dwp用巧妙的方法克服了这些问题。

在泛亚餐厅内，天花板使用了复合高漆板来凸显其高度。背光橙色玻璃和天然玉石装饰突出了酒店的情调，而图形的设计则修饰了空间并为其增添了动感。墙壁经过精挑细选并使用著橙色的黄貂鱼图案装饰，以此来表现墙壁的纹理质感。最终，设计打造了一个时尚、现代的餐厅和酒吧，顾客既可以在此享受轻松的午餐时光又可以度过凉爽的晚餐时刻。dwp为阿尔哈姆拉餐厅设计了一个长长的果汁吧，它是该餐厅的主要特色，闪耀着从蓝色到洋红色的各种色彩。特殊的天花板吊灯照明器具更凸显了这一特色，这些灯具也是由dwp设计的，制造于土耳其。墙壁上有纹理的石头按阿拉伯传统图案堆砌起来，并以黄铜灯做装饰。设计使这里成为了一个拥有明亮、精致外观的现代餐厅。

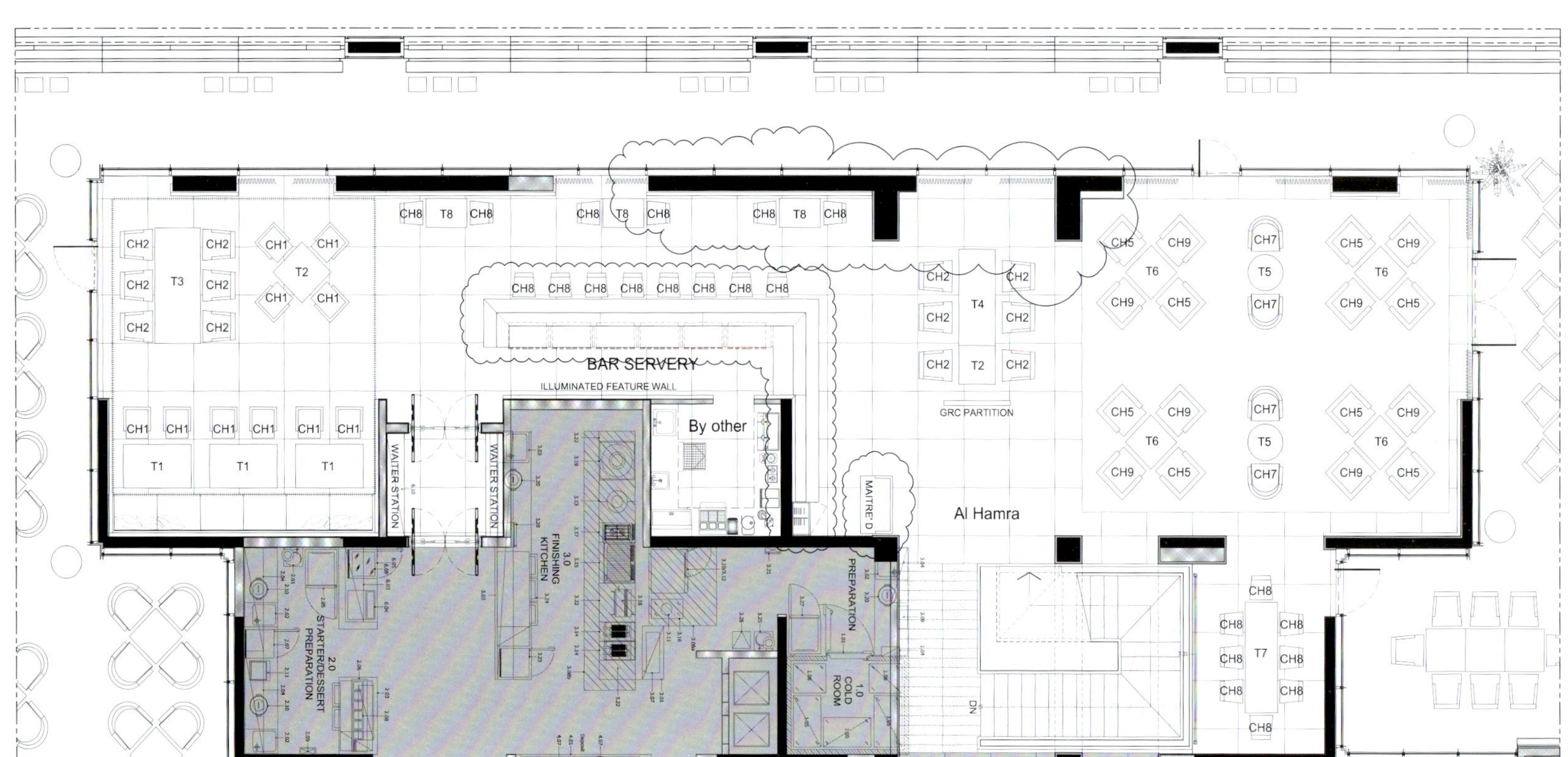
BAR SERVERY
ILLUMINATED FEATURE WALL
By other
GRC PARTITION
Al Hamra
WAITER STATION
WAITER STATION
MAITRE'D
3.0 FINISHING KITCHEN
2.0 STARTER/DESSERT PREPARATION
1.0 COLD ROOM
PREPARATION
DN
T1
T2
T3
T4
T5
T6
T7
T8
CH1
CH2
CH5
CH7
CH8
CH9

Nikken Space Design

Soubu Country Club

Architect
Nikken Sekkei

Interior Designer
Nikken Space Design, Nobuko Suzuki, Yudai Watanabe

Client
Pacific Golf Management

Location
Chiba, Japan

Photographer
Seigan Kuroda

Soubu Country Club boasts a world-class championship golf course that represents Japan in the history of golf tournaments. The ageing building has been transformed into a brand-new clubhouse. The stately building blends in with the natural environment and offers a warm welcome to the guests. Drawing inspiration from the idea of a "residence", the interior retains a human scale and is decorated with different textures and natural materials. Throughout the clubhouse there are various items referencing the history of the country club.

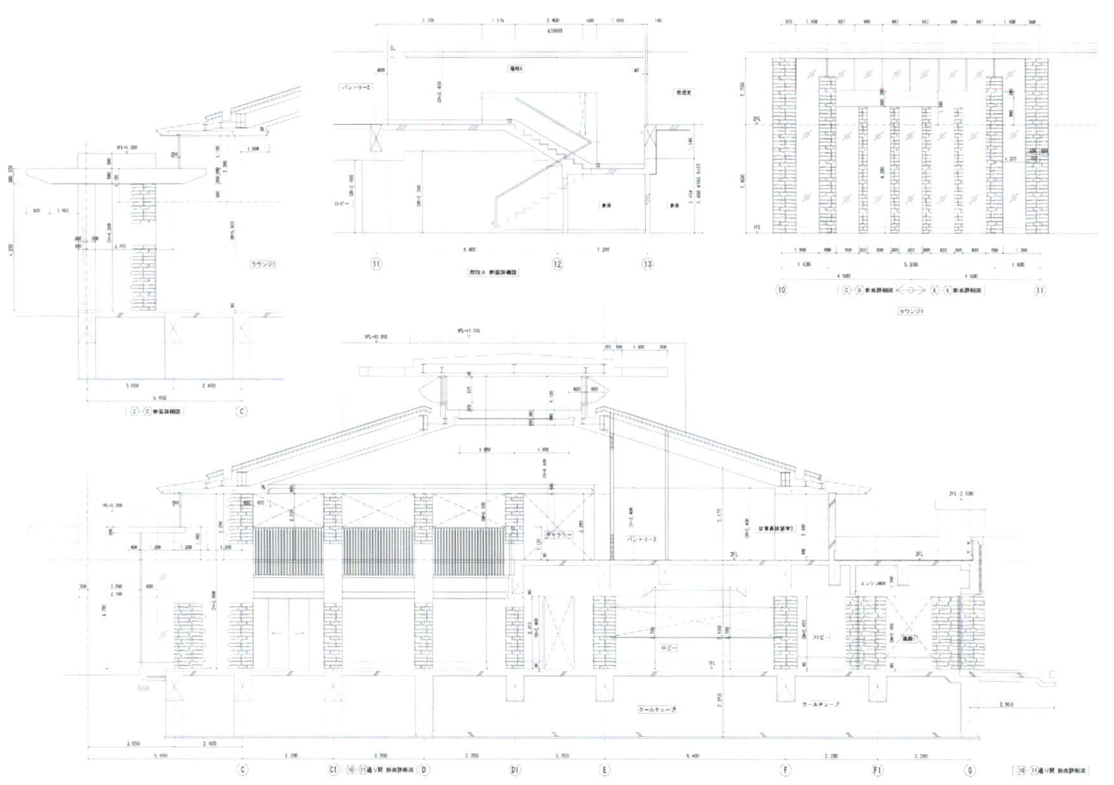
ギャラリー
パントリー2
ロビー
クールチューブ
2FL
1FL
2,960
C
C1
D
D1
E
F
F1
G

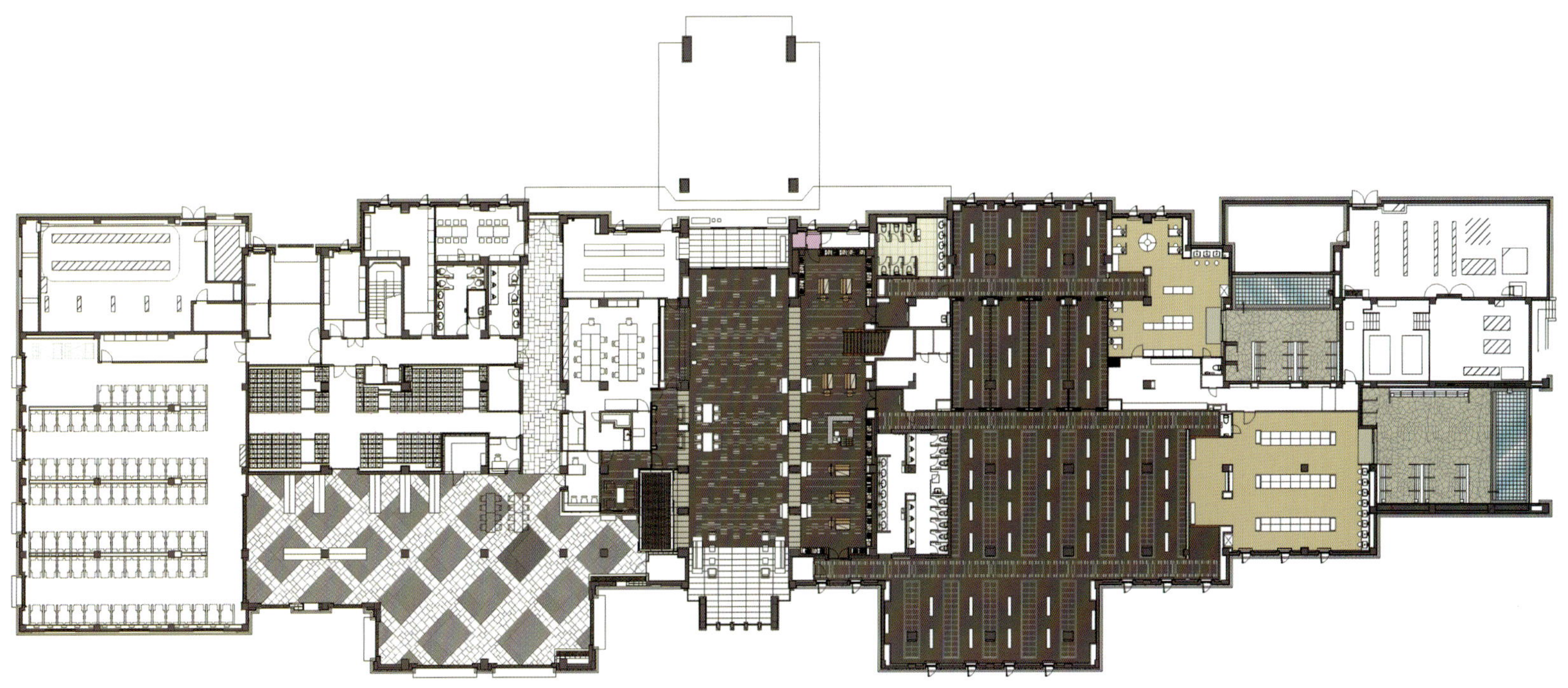

D 600
SW 500
40
30
レストラン チェア
500
450
クッション：本革張り
90
SW 600
90
W 780
60
SW 540
W 660
ラウンジチェア
アームチェア

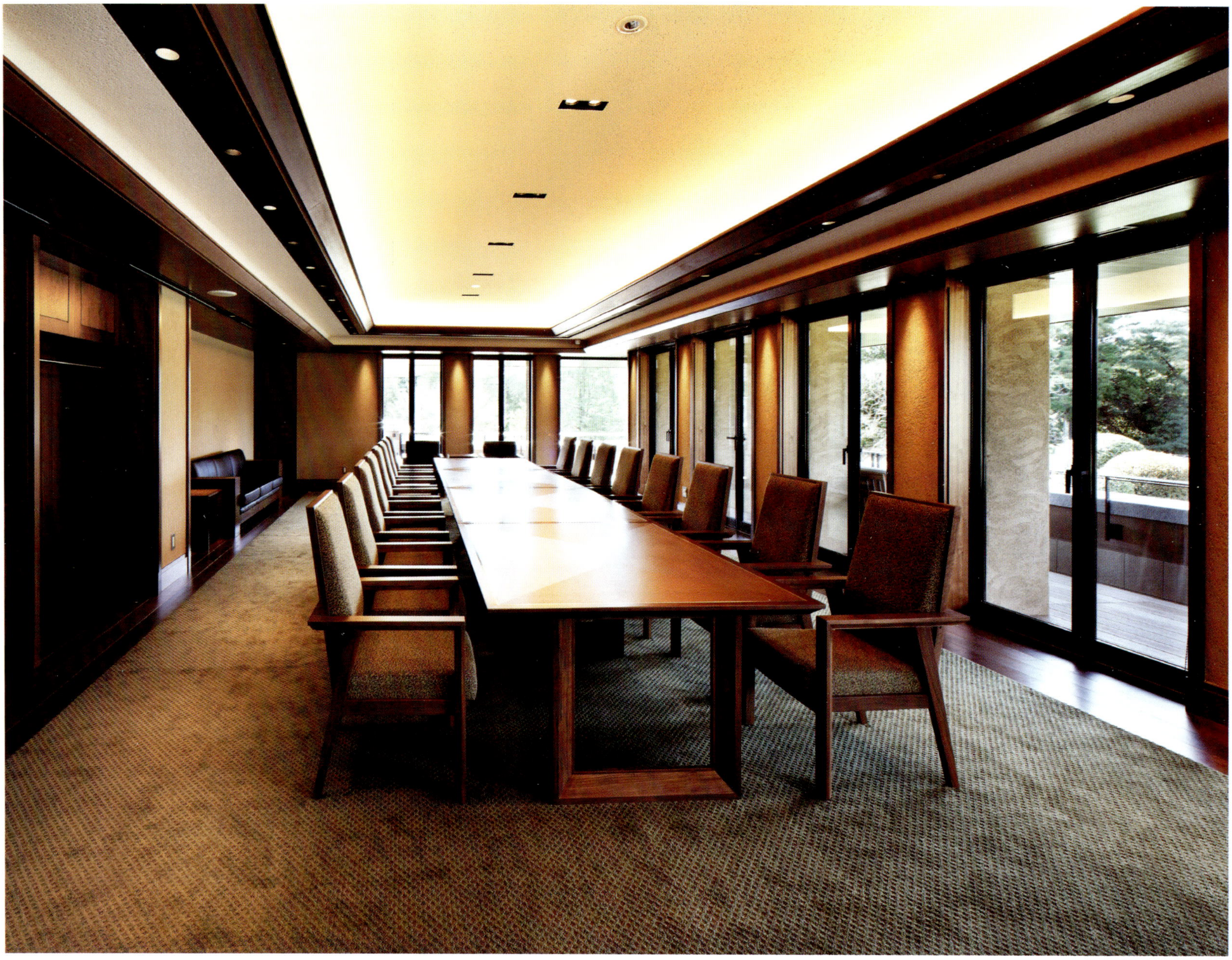

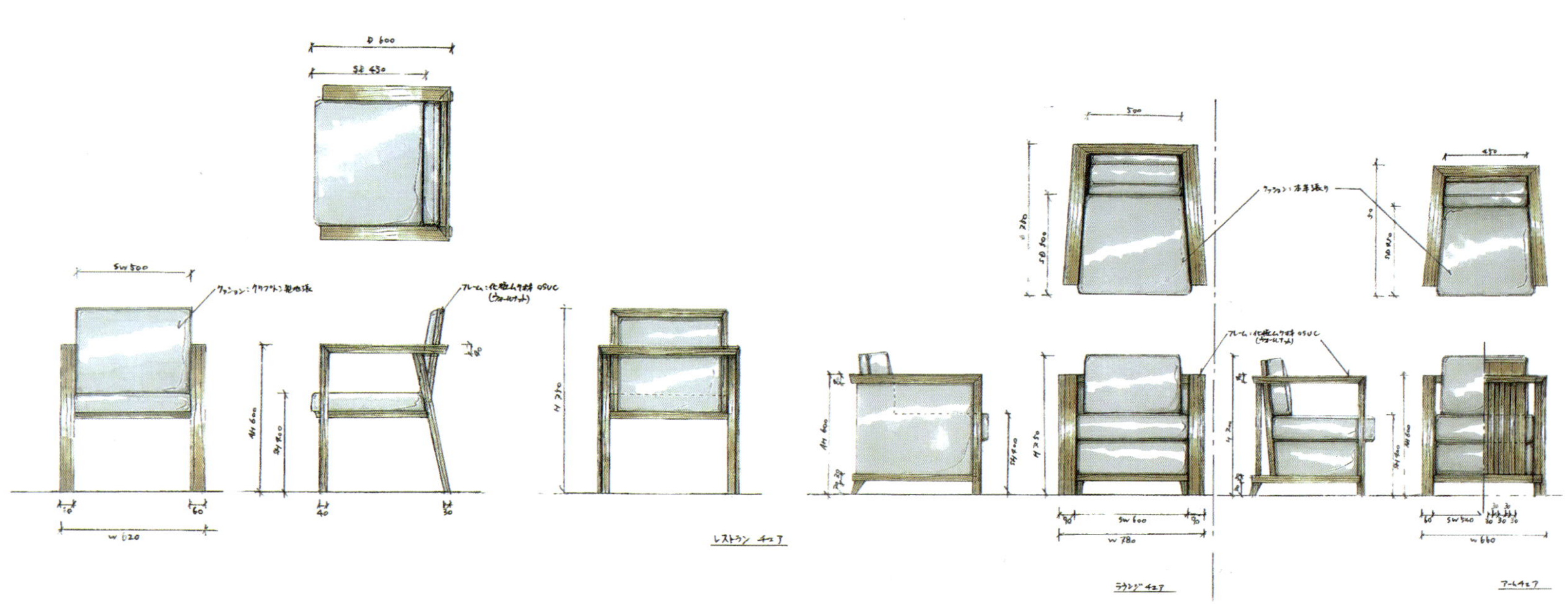

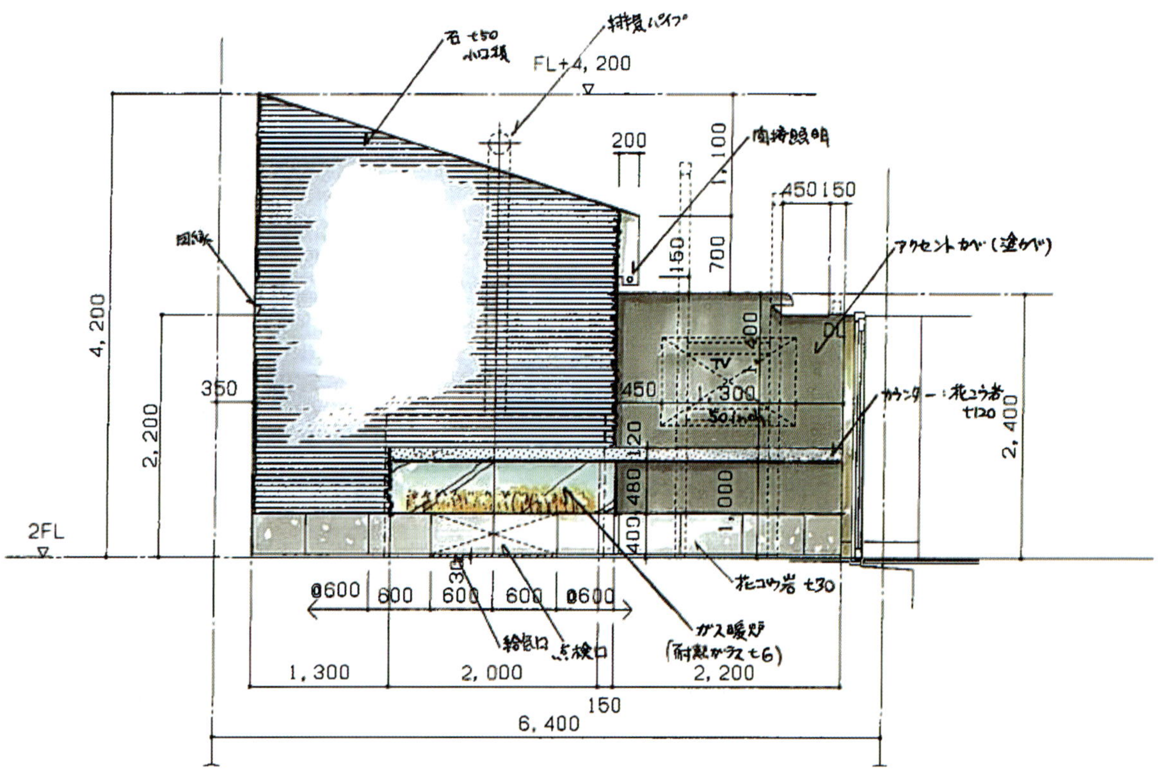

石 t50
小口積
排気パイプ
FL+4, 200
200
100
間接照明
450
150
アクセントかべ（塗かべ）
700
150
TV
400
1, 300
50inch
450
120
カウンター：花コウ岩 t120
480
1, 000
400
4, 200
2, 200
350
2, 400
2FL
花コウ岩 t30
@600
600
600
600
@600
給気口
点検口
ガス暖炉
（耐熱ガラス t6）
1, 300
2, 000
2, 200
150
6, 400

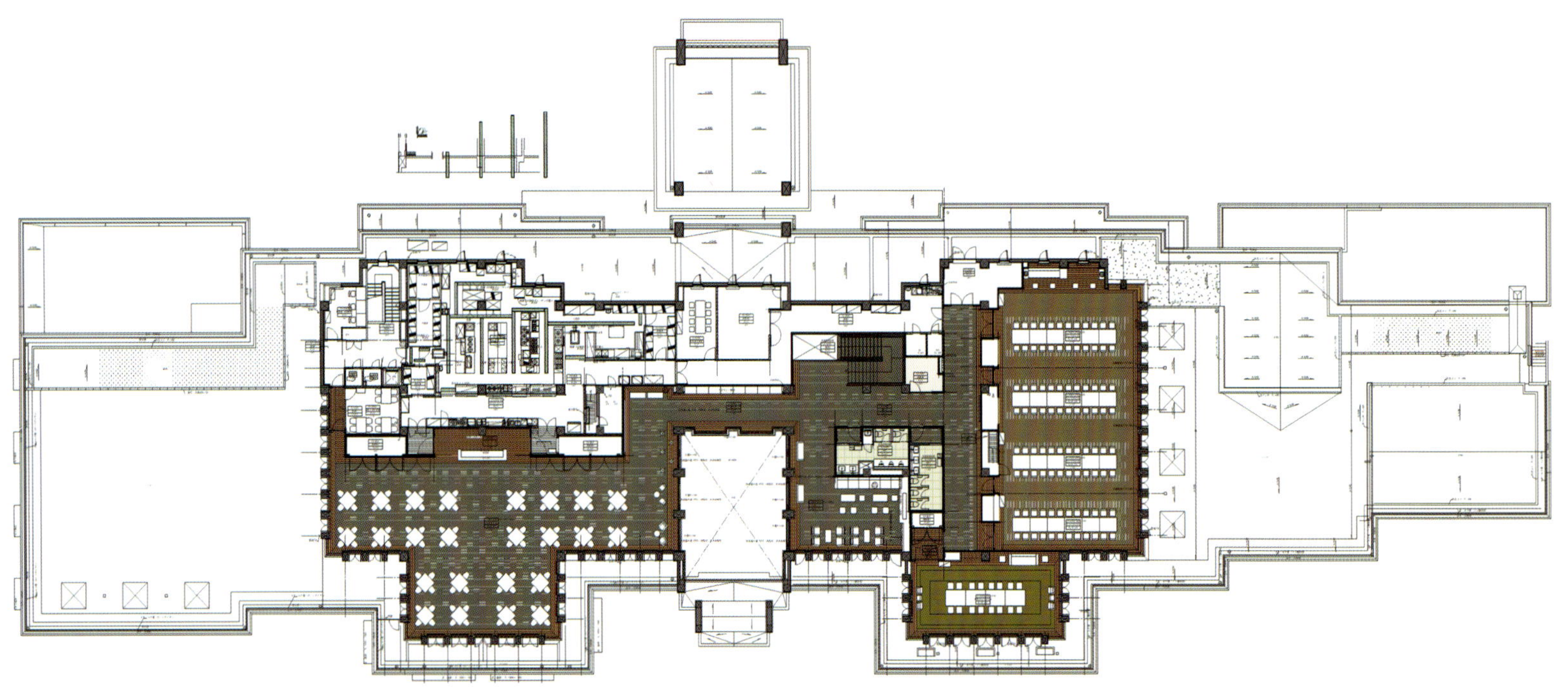

Soubu 乡村会所因拥有世界一流的锦标赛高尔夫球场而自豪，该球场代表了日本高尔夫锦标赛的发展历程。原旧建筑已经被改造成了一个全新的会所。这个宏伟的建筑与自然环境和谐相融，为客人提供了温馨的氛围。其设计灵感源于“住宅”这一理念，室内延续了人文设计这一主线，并装饰着不同的材质和自然材料。遍布于会所各处的摆设，很容易就会让人联想起乡村会所的历史。

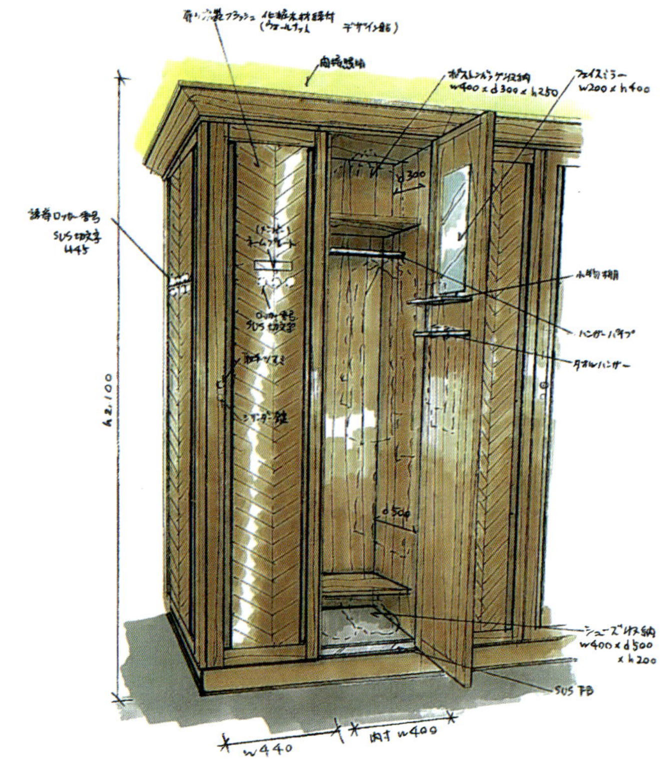

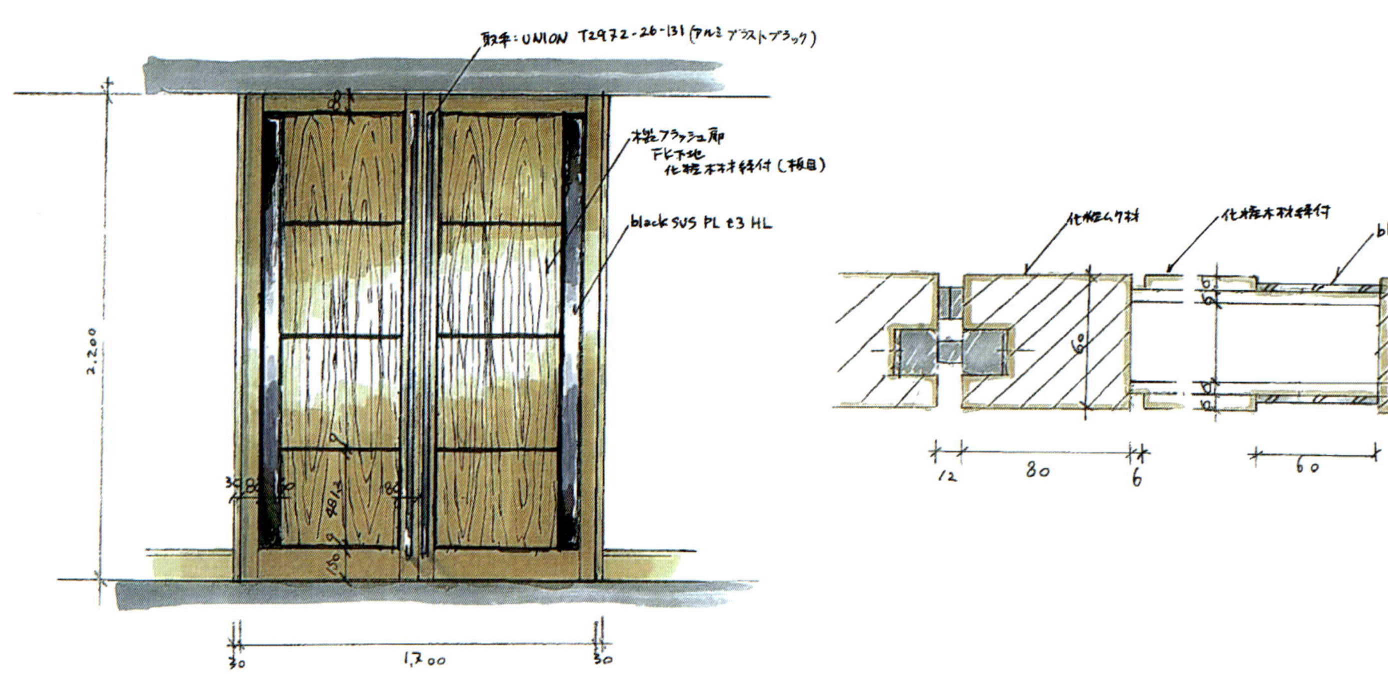

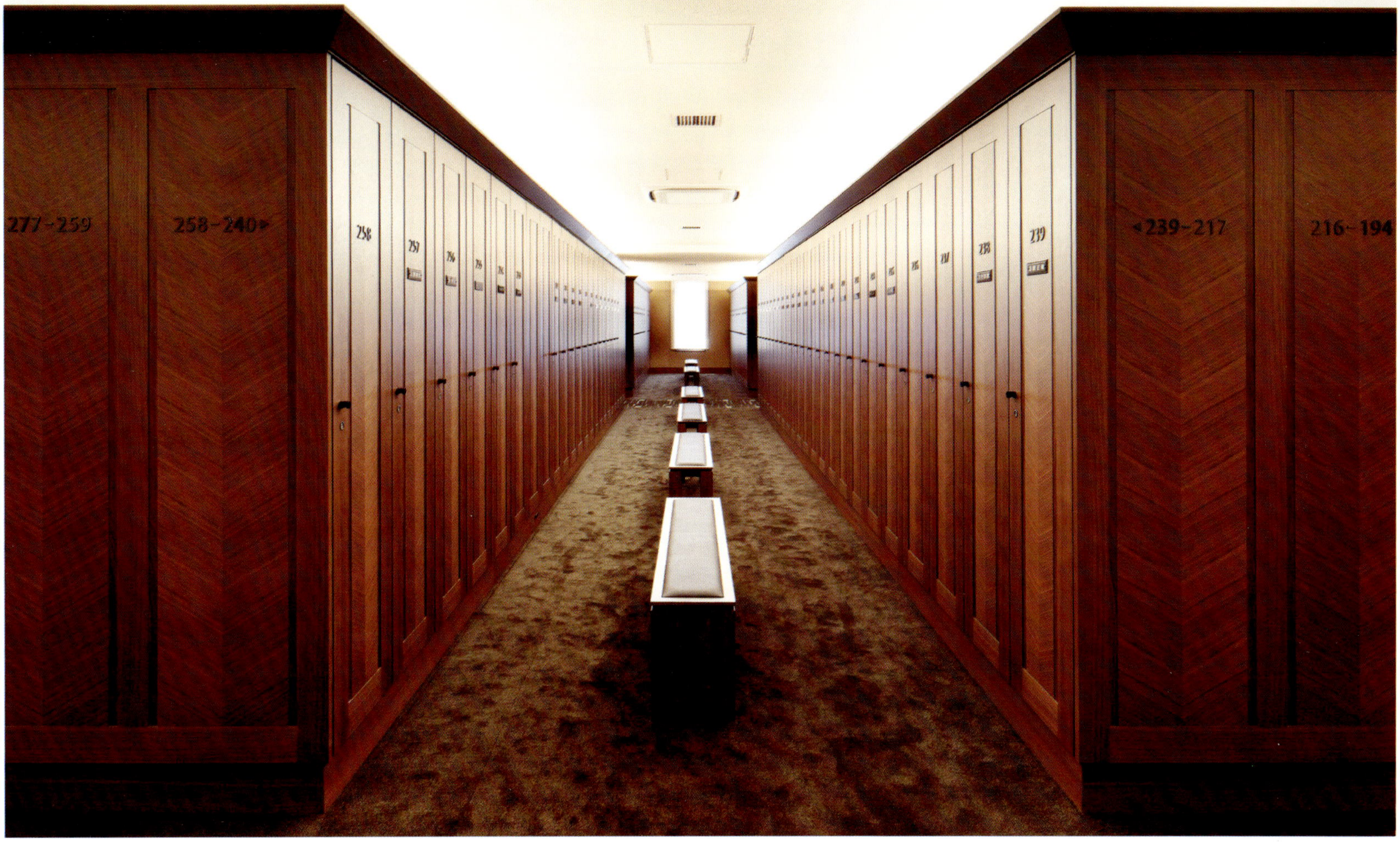
277~259
258~240▸
258
239
◂239~217
216~194

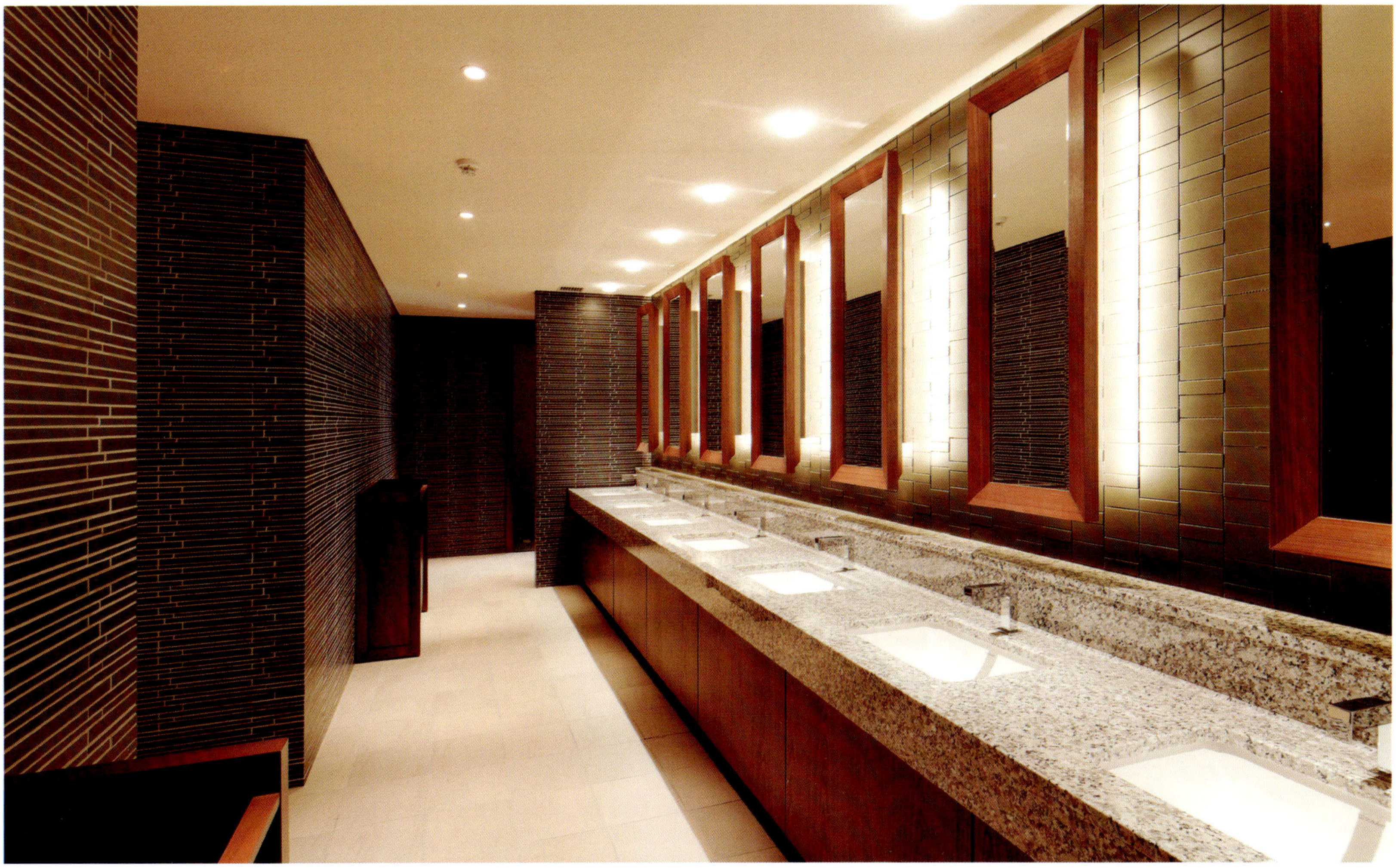

Nikken Space Design

Pines Golf Club

Interior Designer
Nikken Space Design, Takeshi Nakano, Tomoyuki Hino

Location
Aichi,Japan

Photographer
Minoru Karamatsu

This project involved the refurbishment of a 25-year old golf club house.

A Modern-Japanese style was the key concept for this project; the spaces were created with wooden louvers, black stainless steel and traditional Japanese patterned carpets. Each area, including the restaurant, has it's own entrance, which leads directly to the golf course.

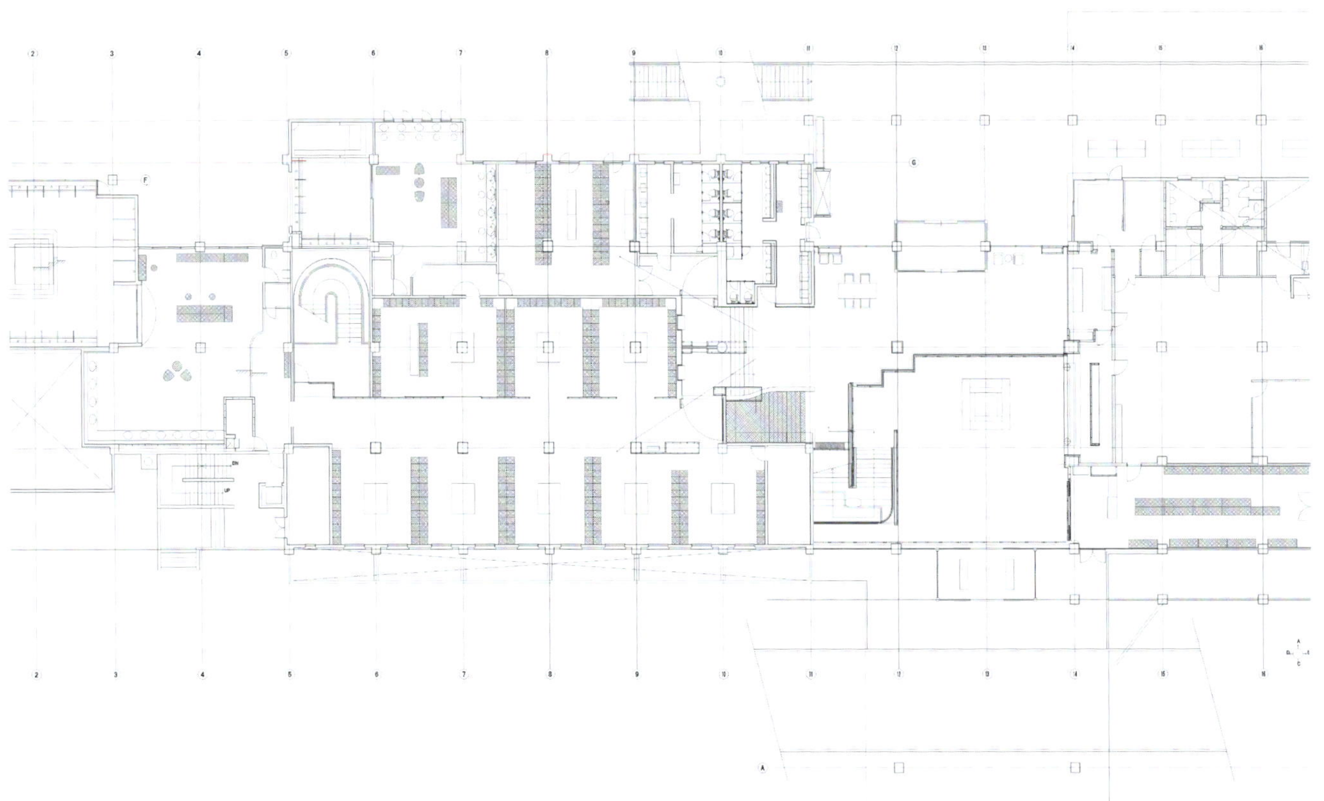

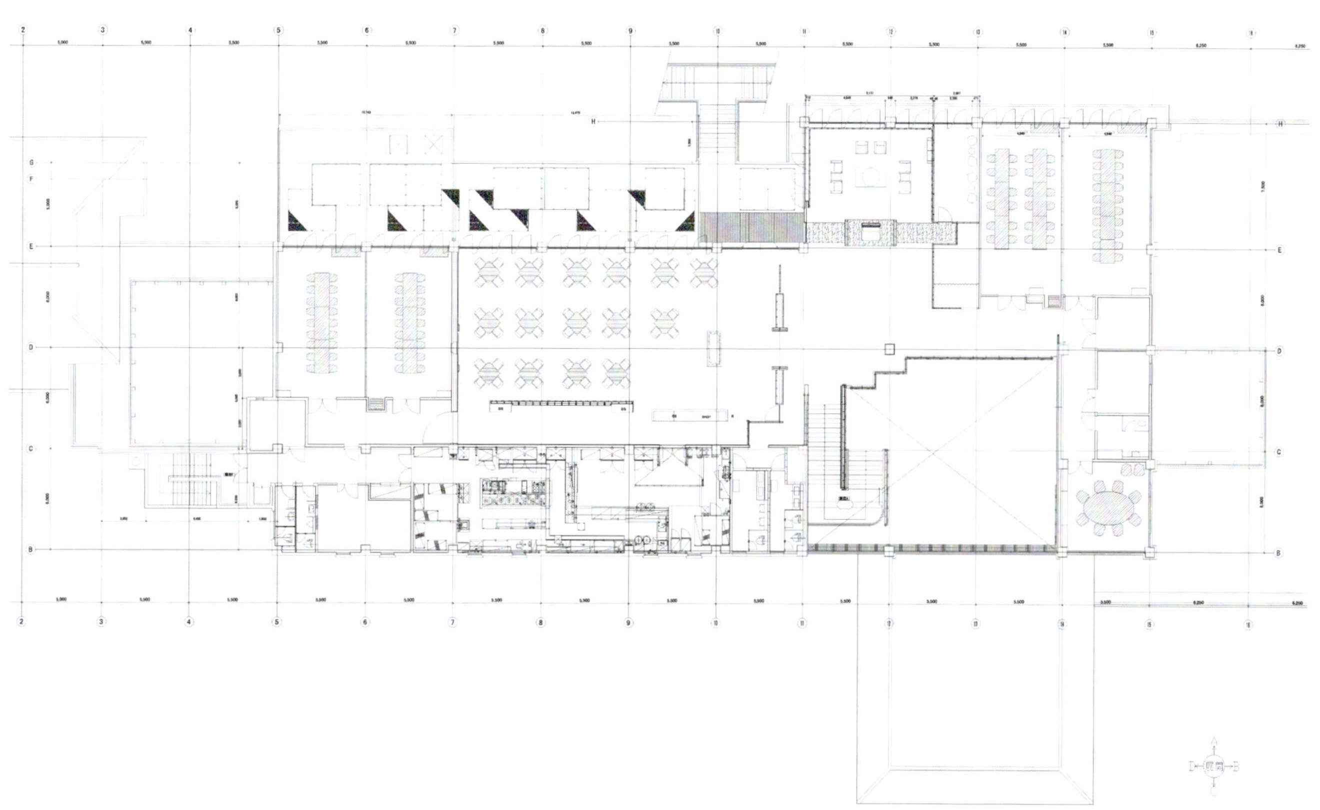

本项目是翻修一个已有 25 年历史的高尔夫会所建筑。

项目以现代日式风格为理念，每个空间都由木制百叶窗、黑色不锈钢以及传统日式图案的地毯构成。包括餐厅在内的每个区域，都有其独立的入口，每个入口都直接通向高尔夫球场。

Christina Zerva Architects

Cartel Club Restaurant

The architect's purpose was to create a sense of a non-dimensional space. Self-illuminating objects gave the dramatic lighting that created the impressive atmosphere and the sense of endlessness. Raw concrete and acrylic glass along with the white interior provides an exciting contrast.

Through the inner space appears a small provocative attempt, in the form of sculptures, silently protesting for a better life on our planet. Dedicated to ecological design, sustainability, clean air, clean water, salads without pesticides, apples with worms, world without violence and bloodshed, objects hanging from the roof with chains remind us of the consumerist society we live in.

Recycled and second hand materials have been used for the building. For the insulation is used cellulose rather than the common building insulation materials that may contain toxics. Also photovoltaic solar panels provide sustainable electricity for any use.

The entire Club is illuminated with LED modules that base their success on their characteristic properties such as low energy consumption, zero ultra-violet and infra-red radiation, long life and extremely compact dimensions. They offer greater performance, more lumens per watt and therefore are more eco friendly.

Cartel provides a capacity of over 2,000 people. It is a steel structure with a flooring of polished concrete. The roof opens in the summer and closes when cold.

It is separated in different levels. The two lounges provide intimacy. In the middle of the space there is a raised section where the DJ's turntables are. The extraordinary lightning options and the capability of the technical and electrical operative systems have attracted many of the top DJ's that have guested in the Club.

The exterior is painted black in contrast to the interior. The white inner walls are used as giant video projection surfaces changing the character of the Club whenever needed.

Leading Engineer
Milos Savic

Area
1,200 m^2

Lighting Designer
Dimitris Nasikas

Photographer
Mihajlo Savic

Location
Larissa, Greece

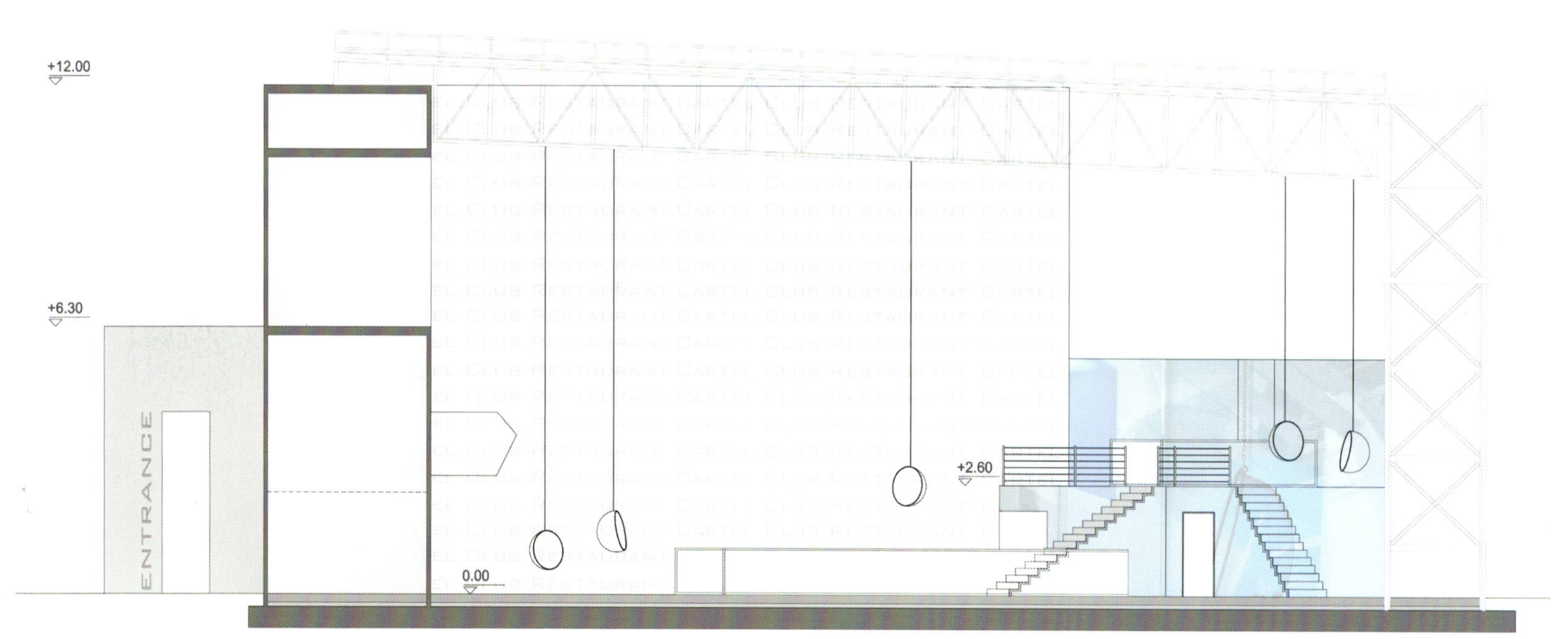
+12.00
+6.30
ENTRANCE
CLUB RESTAURANT CARTEL
+2.60
0.00

+12.00
+2.60
+2.40
+2.40
0.00
0.00
0.00

BAR RESTAURANT
RAMP DOWN
ENTRANCE
ENTRANCE
+2.60
+2.40
+2.40
+-0.00
+12.00
+2.40
6.84
5.6594
1.1267
3.6678
7.4328
5.1424
7.4819
3.8428
4.6349
2.7296
4.1064
2.0058
3.2766
6.7952
16.9811
7.5499
3.3718
5.7624
4.1256
1.1931
2.2419
1.9769
12.7778
12.3983
2.0365
6.6580
4.8543
1.4264
1.3977
3.6912
5.7976
4.8325
3.6295
7.0298
4.7919
0.8789
2.9490
4.3249
20.2
12.3863
3.9000
23.2
4.1000
39.9801

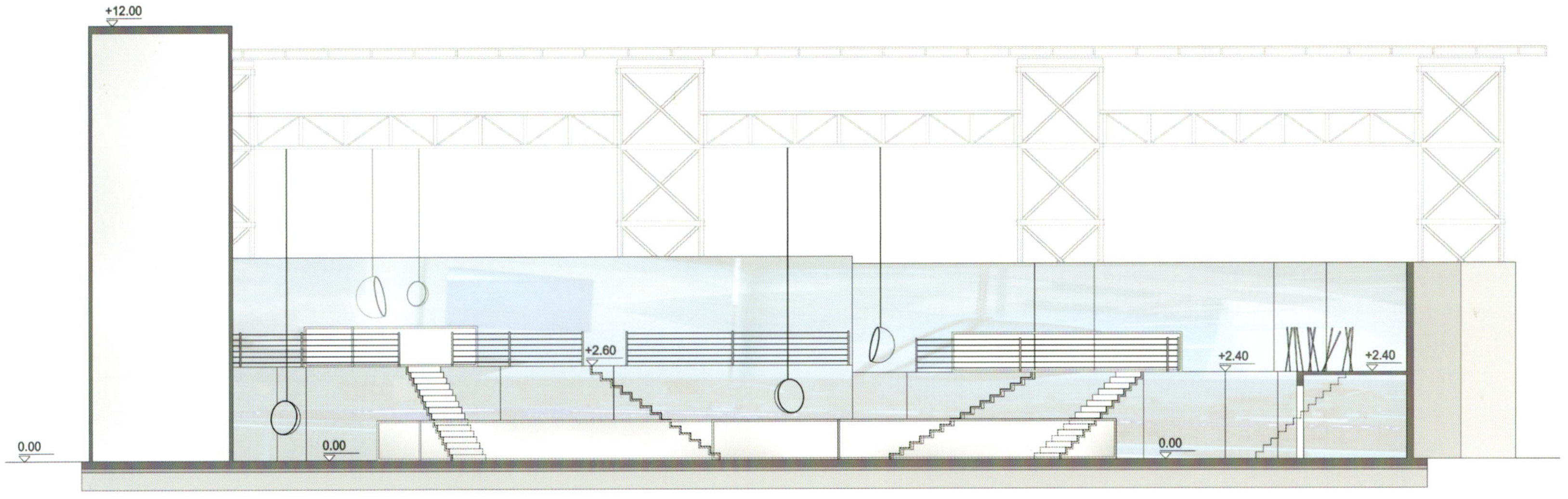

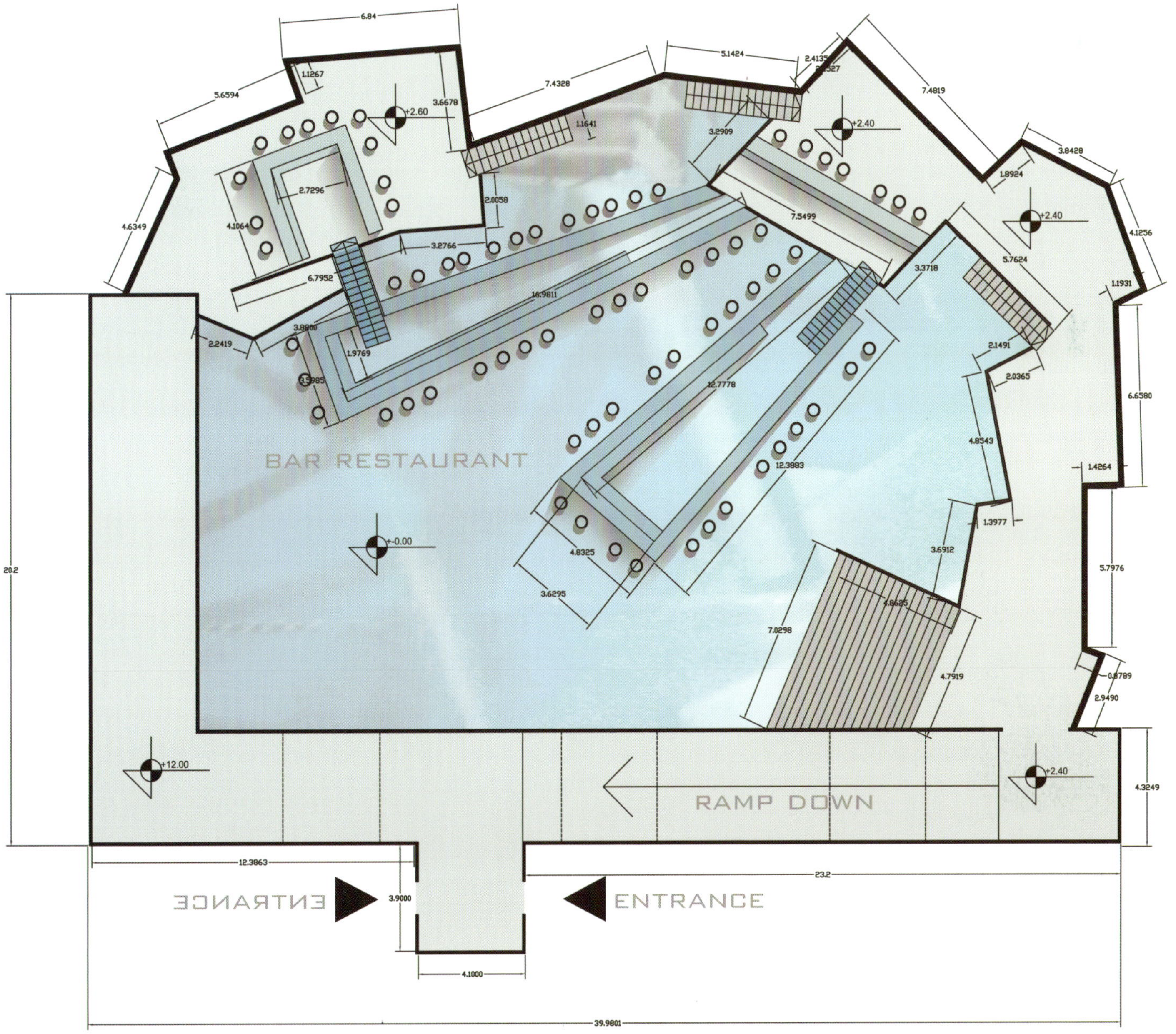

设计师的目的是创建一种无量纲的空间感觉。自身可发光的物体带来了绝佳的照明效果，创造了令人印象深刻的气氛和无穷无尽的感觉。混凝土原材料、丙烯酸玻璃与白色的室内设计产生了一种令人震撼的对比效果。

内部空间的设计展示了一次小小的突破性的尝试，以雕塑的形式呈现，体现着想要更好地生活在这个星球上的努力。设计师致力于生态设计、可持续发展、清洁空气、清洁水源、无农药沙拉、有虫苹果、没有暴力和流血事件的世界等主题；而那些用铁链悬挂在屋顶上的物体则提醒着我们，我们正生活在消费型社会中。

建筑设计中使用了可再生的和二手的材料，同时设计还使用了纤维板绝缘材料，用以代替那些可能含有有毒物质的常见的建筑绝缘材料。此外，太阳能光伏电板可为项目提供可持续的电力。

整个会所都使用 LED 模块照明，因其在低能耗、零紫外线和红外线辐射、使用周期长和紧凑节约型特点上优势突出。LED 模块也使会所更具表现力、更省电，也因此，更加生态友好、节能环保。

Cartel 能够容纳 2000 多人，为钢结构建筑，地板由抛光混凝土制成。建筑屋顶可在夏天打开，在寒冷时关闭。

不同的区域相互分离，两个休息室提供了私密空间。在空间的中部有一个凸起的部分，那里便是 DJ 台。多种绝佳的闪光效果、高端的技术设备及电气操作系统已经吸引了很多顶尖的 DJ 来这家会所演出。

室外被漆成黑色，与室内形成鲜明的对比。白色内墙被当成巨大的视频投影面，可根据需要随时改变会所的特征。

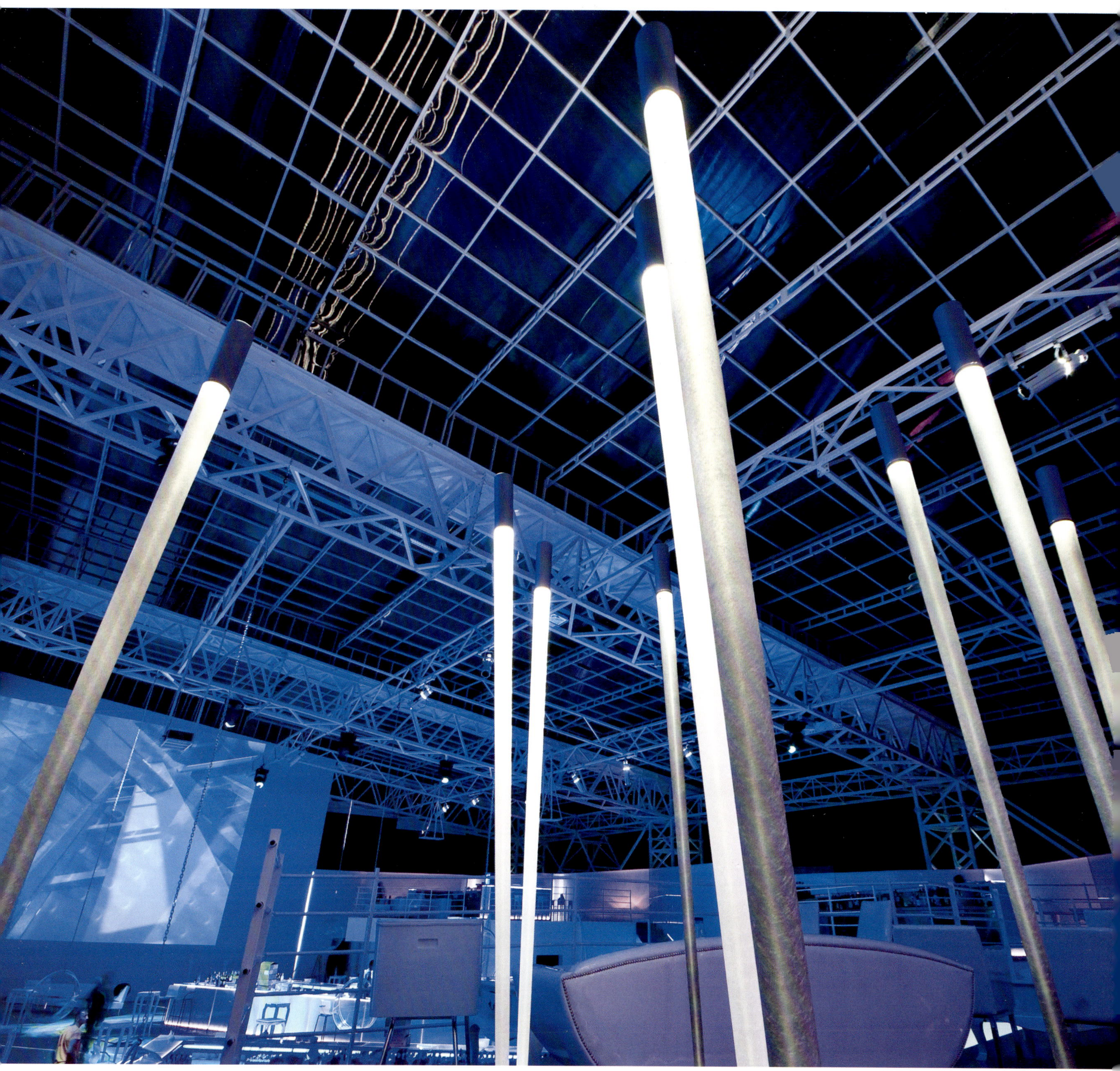

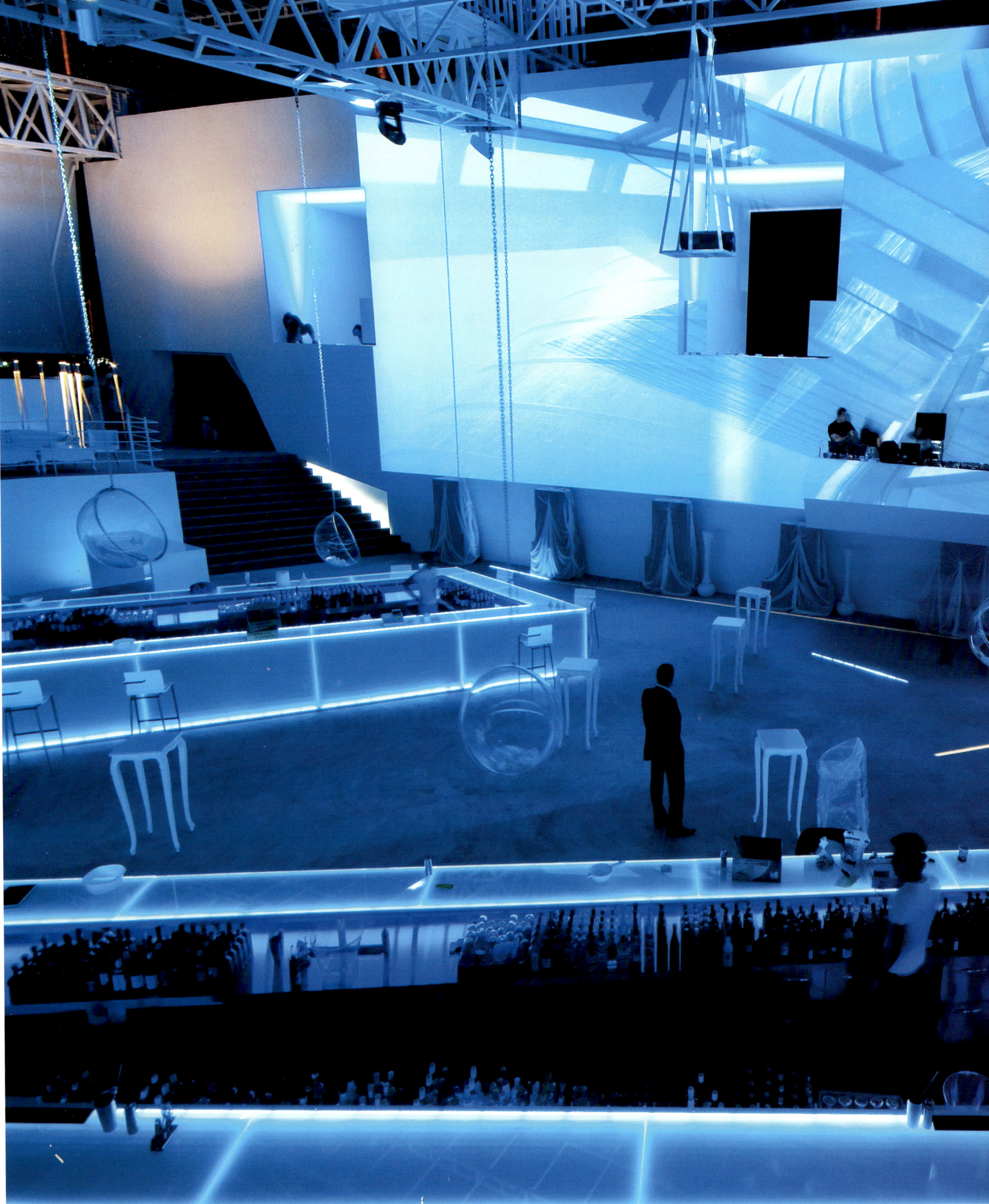

夜总会 NIGHTCLUB

Albert Angel

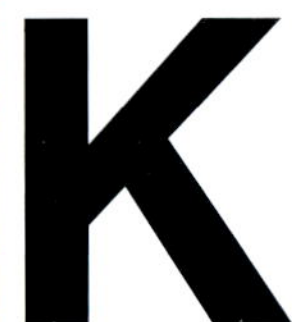

ONOBA Restaurant & Bar

Project Team
Albert Angel, Jacques Hanekom, Jonathan Legrand

Construction Company
Green Ocean Investments, Pty, Ltd. Seychelles

M&E Engineer
Eme Overseas, Mauritius Seychelles

Landscaping
Albert Angel

Location
Angelfish Bayside Marina, Roche Caiman, Mahe, Seychelles

Area
1,200 m²

Design Brief

The brief was to create a new type of "social destination" for the island of Mahe – a space that transforms from family restaurant to night club at a flick of a switch – catering to a wide audience from local residents to itinerant tourists.

The space had to be colorful and approachable yet sleek and modern. It had to address the client's love of boats as well as respond to its marina surroundings.

Design Challenges

The challenge was to create and execute a design with almost non-existent local materials and scarce building and crafting skills. Its building scheme had to be simple. A high level of salinity was to be considered in the choice of materials and furniture. Tropical heat and strong easterly winds determined the placement of various zones of the establishment.

Design Results

The result is an interior that encapsulates its marina surroundings and celebrates the underwater life of the Seychelles. A sculptural shoal of 8,000 stainless steel fish-like pieces leads you from the entry to the main space where it culminates in a massive swirling element hanging from the coral inspired decorative beams above the bar. With detailed lighting designs, the entire space changes from daytime resort chic to a vibrant bar scene.

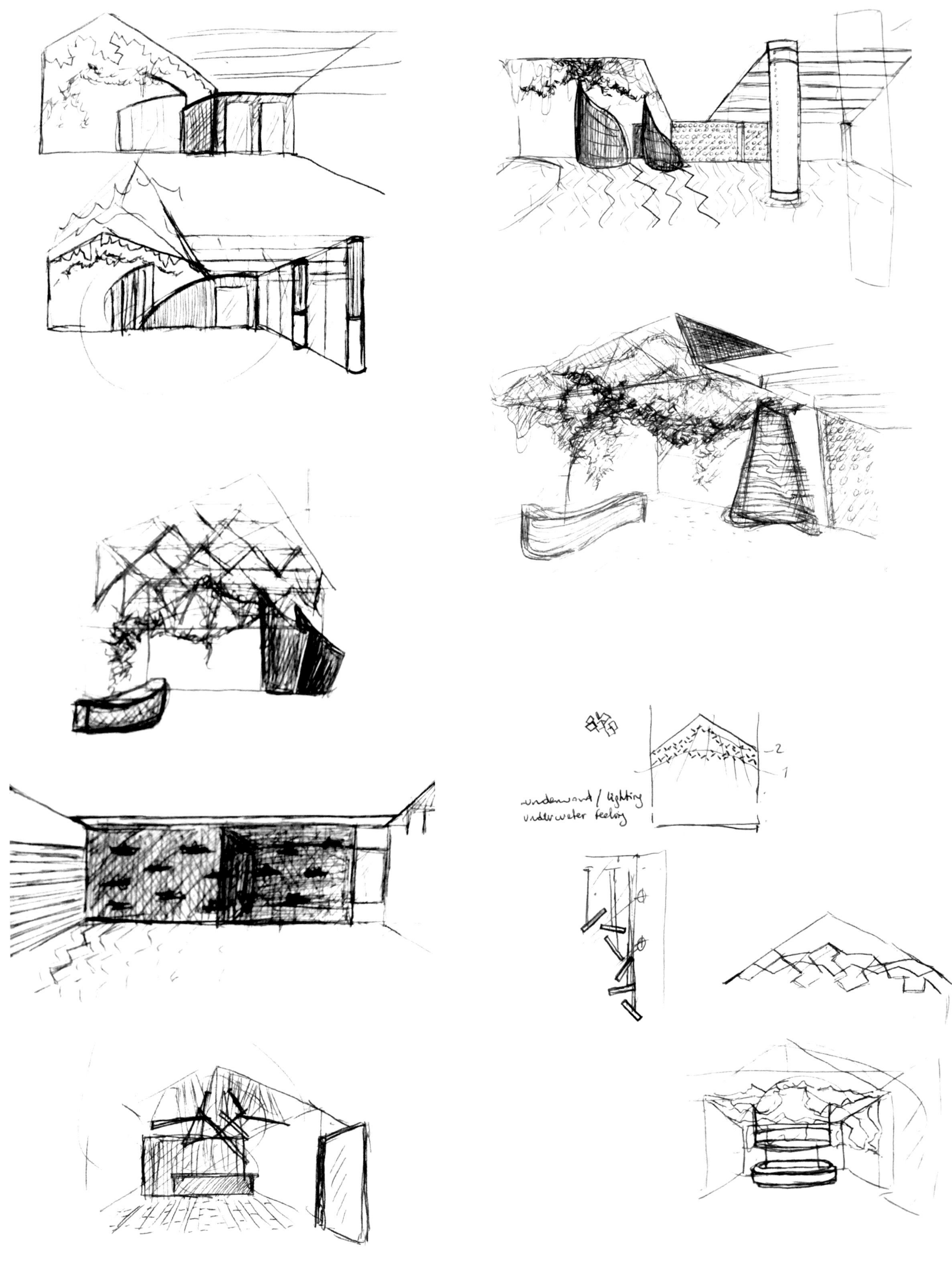
2
1
underward / lighting
underwater feeling

设计概要

本案旨在为马埃岛开设一个新型的社交场所。在这个空间里，家庭餐厅转瞬之间就能变成夜总会，满足包括本地居民和流动游客在内的广大顾客的需要。

本案空间色彩丰富，设计友好亲切而又不失时尚感和现代感，既满足了顾客对船只的喜爱，又与周围的海港环境相呼应。

设计挑战

挑战在于需要使用当地几乎不存在的材料和罕见的建筑工艺技能来创建并执行设计。本案建造方案必然要简洁。在选择材料和家具时，高含盐度是一个重要的考量因素；而热带高温和强东风则决定了各个分区的布局模式。

设计成果

本案的内部设计如周围海滨环境的缩影，颂扬了塞舌尔的水下生命。入口处是一个由 8000 个鱼形不锈钢部件组成的鱼群雕塑，引导顾客进入主厅，雕塑一直延伸到一个巨大的旋涡状物体（它位于酒吧上方，悬挂在一个受珊瑚启发而设计的装饰性横梁上）上。复杂精细的照明设计可使本案的整体空间从白天漂亮的度假胜地过渡到充满活力的酒吧场景。

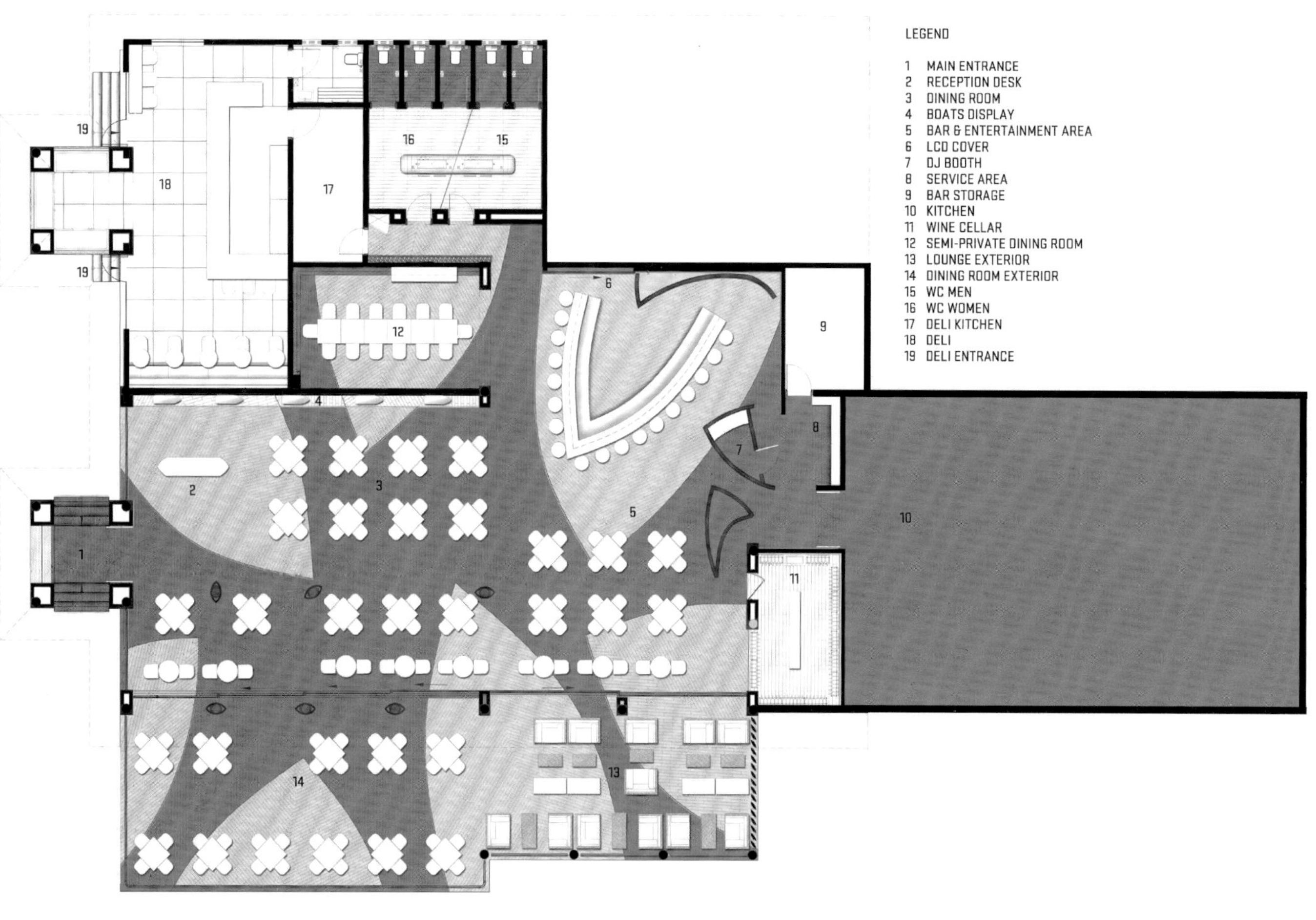
LEGEND
1 MAIN ENTRANCE
2 RECEPTION DESK
3 DINING ROOM
4 BOATS DISPLAY
5 BAR & ENTERTAINMENT AREA
6 LCD COVER
7 DJ BOOTH
8 SERVICE AREA
9 BAR STORAGE
10 KITCHEN
11 WINE CELLAR
12 SEMI-PRIVATE DINING ROOM
13 LOUNGE EXTERIOR
14 DINING ROOM EXTERIOR
15 WC MEN
16 WC WOMEN
17 DELI KITCHEN
18 DELI
19 DELI ENTRANCE

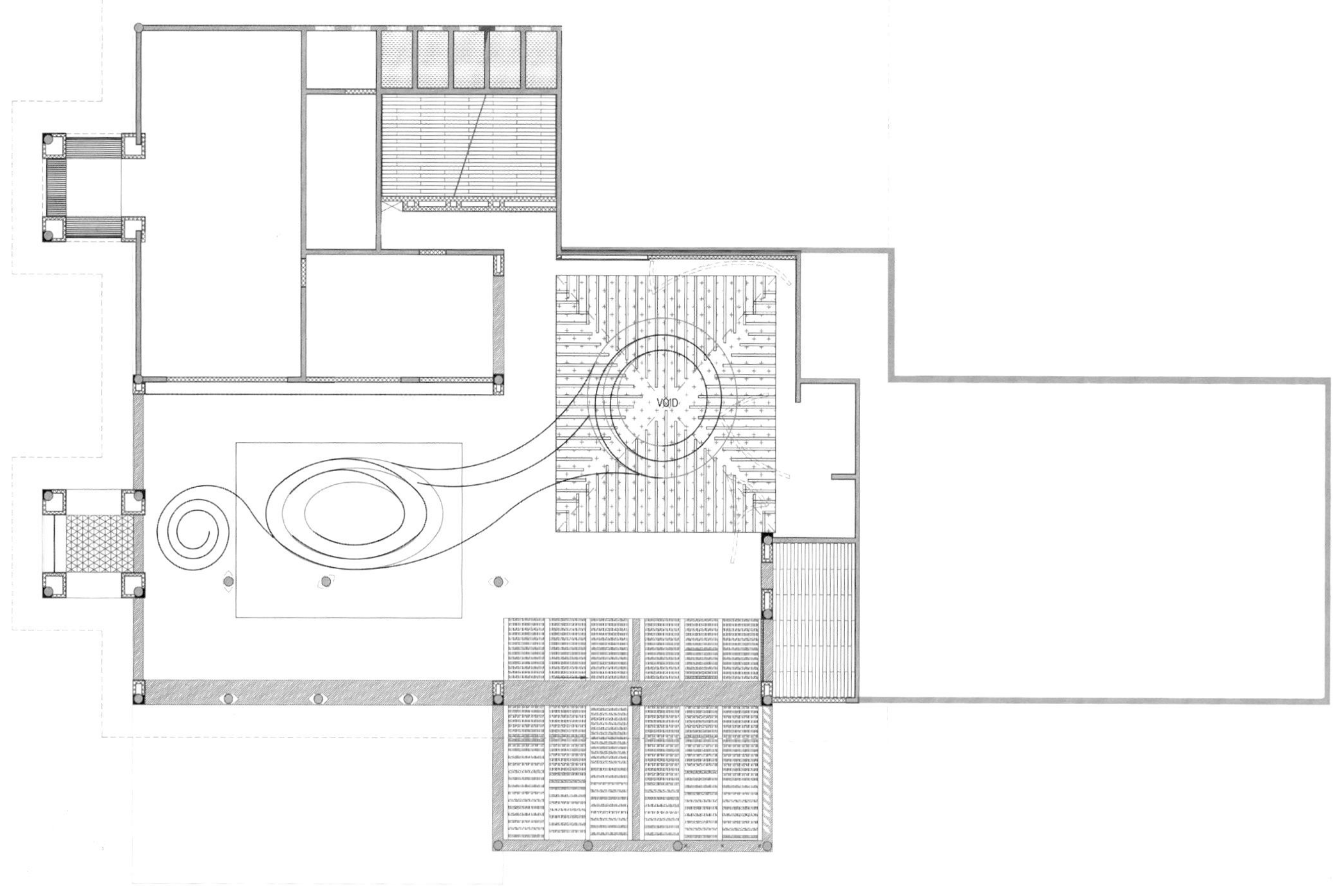
VOID

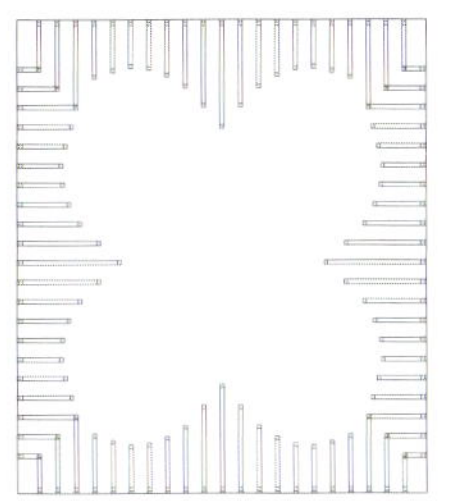
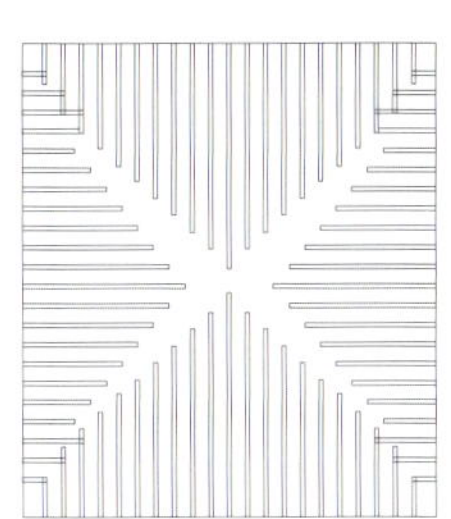
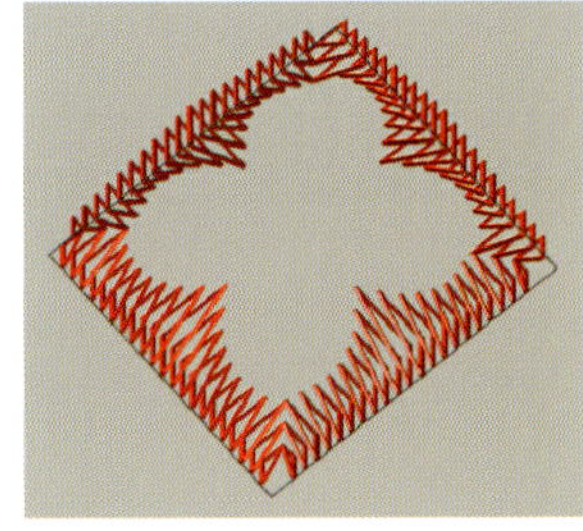
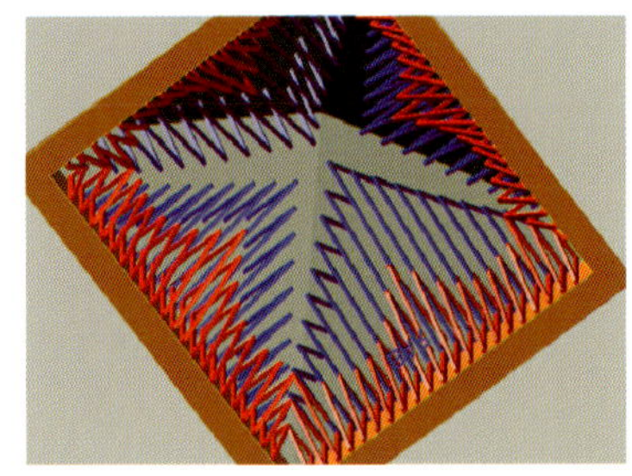

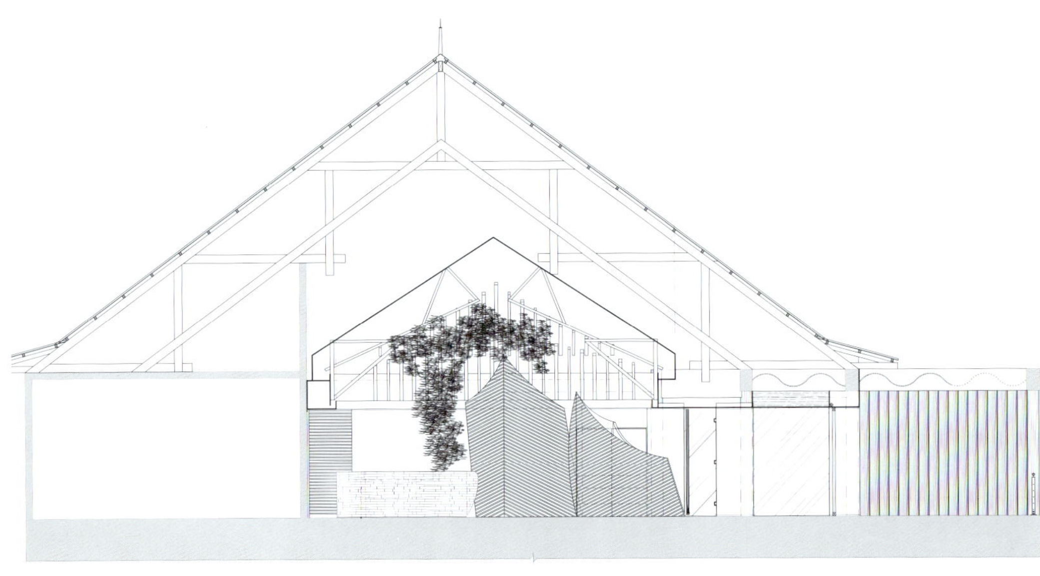

Orbit Design Studio

Location
Yas Island, Abu Dhabi

Area
1000 m^2

Photographer
Owen Raggett

Allure is the first nightclub by legendary restaurateurs Cipriani. Bearing in mind the pedigree of Cipriani, its location on the exclusive Yas Island marina development and placement overlooking the Formula 1 race track, this was a project that had to appeal to the highest end of the international luxury crowd.

Orbit met and exceeded all expectations with a striking design that combines sensual, nautical inspired seating with fractal walls and ceiling complete with infinite color control in The Main Room. Pink gold leaf and distinctive bronze cladding add another level of glamour.

The Terrace overlooks the race track and sports VIP tables and PODs for those who really want to indulge. The final touch for VIPs is their own elevator. All this has combined to put Allure firmly in at the top level of luxury clubs.

Allure 是由具有传奇色彩的餐馆老板 Cipriani 成立的第一个夜总会。考虑到 Cipriani 的背景，夜总会位于独特的亚斯岛码头开发区，从这里可俯瞰 F1 赛道，该项目是为高端的国际奢侈品人群量身打造的。

Orbit 设计公司提交了一个精彩绝伦的设计，其效果满足并超越了原有的预期，该设计将感官效果极佳、具有航海感觉的座椅与不规则的墙面和天花结合在一起，并在主要空间内运用了多种灯效。粉色的金箔和独特的青铜包层又为空间添加了一丝不同的魅力。

在阳台上可俯瞰到赛道、体育贵宾桌以及为那些想尽情狂欢的人们而准备的前排座椅。该夜总会设计的点睛之笔是宾贵们的专用电梯。所有的这一切都让 Allure 稳居顶级豪华夜总会之列。

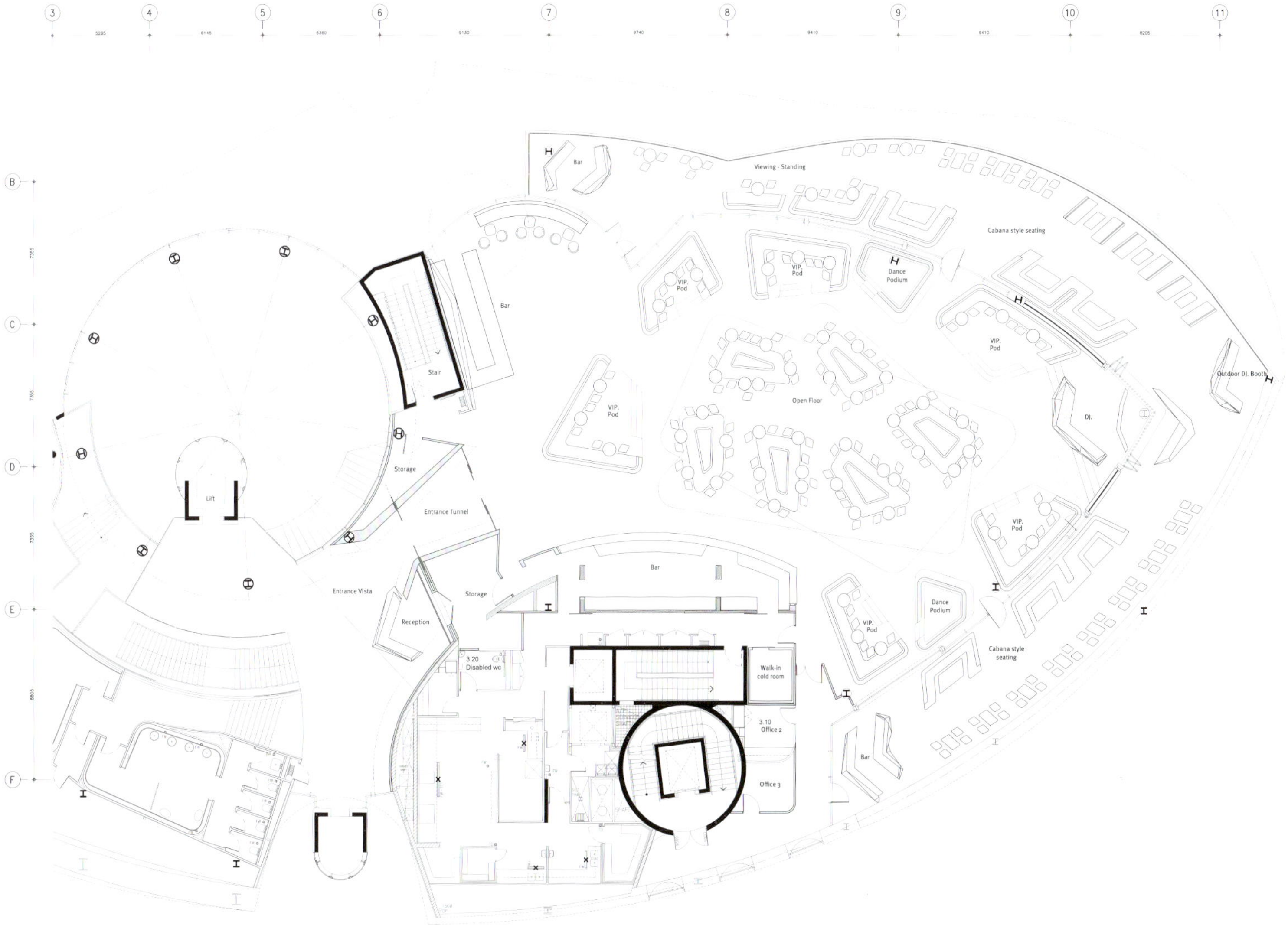

3 4 5 6 7 8 9 10 11

B C D E F

Bar
Viewing - Standing
Cabana style seating
VIP. Pod
Dance Podium
Bar
Stair
Open Floor
Outdoor DJ. Booth
DJ.
Storage
Entrance Tunnel
Lift
Entrance Vista
Reception
Storage
Bar
Dance Podium
Cabana style seating
3.20 Disabled wc
Walk-in cold room
3.10 Office 2
Office 3
Bar

Parolio & Euphoria Lab

Club Musée

Designer
Parolio

Area
400 m^2

Location
Madrid, Spain

The brief was to create a new club and multi-use space for the competitive nights of Madrid with an international and powerful identity. That was the goal set by owners Fernando Nicolas and Jacobo Dominguez and his associetes.

Parolio crated the concept, branding identity and interior design for the project by merging the added cultural value of an art gallery with the exiting and colorful universe of the club scene.

The result is a stimulating mix of streamline interior design, dramatic color structures and art.

All the original furniture design by Parolio has been treated as sculptures with the use of multiple fabrics creating a "color block effect" patchwork.

In the VIP room the sofas are a continuation of the photographs that form a dramatic back drop. All columns and walls are covered in black glass with geometric shapes.

Photography, video installations, and illustration by the designer and renown artists such as Paco Peregrín, Robert Bartholot and Glenn Hillario are on display in the club making it a complete cultural experience.

On the dance floor a floating DJ table is completed by an installation of Acqualed tubes three meters wide and a geometric lamp on the ceiling.

The club has become since its opening in winter 2010 the place to be in the Madrid night scene and a popular venue for cultural and fashion events.

ART YOU ART
ART YOU ART YOU
YOU ART YOU
ART YOU ART
YOU ART YOU
ART YOU ART

这个设计旨在创建一个新型夜总会、一个多用途的空间，以充满国际性和力量感的特征使其在马德里充满竞争性的夜晚中崭露头角。这是由业主 Fernando Nicolas 和 Jacobo Dominguez 以及其合伙人共同制定的目标。

Parolio 通过将艺术画廊增加的文化价值与正在消退的多彩的夜总会场景相结合，为这个项目进行包装概念、品牌身份和室内设计的定位。

最终，该空间是一个集合了流线型室内设计、戏剧性的色彩结构和艺术特色的令人兴奋的混合体。

设计通过使用由多种面料拼接而成的拼布，创造出了“色彩块效应”，这样的效果使得 Parolio 设计的所有原始家具都被视为雕塑艺术品。

在贵宾室内，沙发是照片的延续，形成了一种戏剧性的背景。所有的柱子和墙壁上都布满了几何状的黑色玻璃。

由设计师和著名的艺术家 Paco Peregrín，Robert Bartholot 和 Glenn Hillario 完成的摄影、影像装置和插图都陈列在夜总会内，营造出一种完整的文化体验。

在舞池内，浮动的 DJ 台后是由三米宽的 Acqualed 管子组成的管状墙，天花板上还悬挂着一个几何形灯饰。

自 2010 年冬天开业以来，这家夜总会已经成为马德里夜景不可缺少的一部分，也是流行文化和时尚活动的集散地。

ART
YOU
BEAUTIFUL
ART
YOU

Munge Leung

Dragonfly Nightclub

Designer
Giancarlo Garofalo, Alessandro Munge, Sai Leung, Mehari Manna Seare

Location
Niagara Falls, Canada

Photographer
Tom Arban

The owner of this nightclub was thrilled at the opportunity to redefine the night scene in Niagara Falls Canada, and to be part of The Fallsview Casino Resort retail and entertainment complex. To perpetuate the excitement and invigorate visitors from sunrise to sunset, a new vision for night-time entertainment at the tourist city of Niagara Falls, Canada needed to be imagined. The city's investment in the development of The Fallsview Casino Resort presented an opportunity to create a new wave of Las Vegas-style entertainment venues. The resort's management company felt as though a new nightclub venue would appropriately contribute to the excitement of The Fallsview Casino Resort; hence, the inception of this 1115 m^2 full-service entertainment venue yielding a capacity of approximately 800 people. Apart from first-class nightclub hospitality, this entertainment venue provides the perfect setting for multimedia presentations, corporate events, fashion shows and promotional events.

The functional vision aimed to create a multi-functional entertainment venue; the creative vision was to create a mystical and exotic environment that would attract nightclub goers from around the world. The ultimate goal was to create an entertainment escape, where sounds, lights, people, and décor harmonized to captivate an exhilarating experience transcending time and space. The biggest challenge during the design process was its sound-proofing requirements. The building's developer did not account for a nightclub to be placed on the lowest level of the complex; and so, special design work was required to accommodate a successful operation without disturbing neighboring retailers, hotel guests. In order to incorporate a proper sound-proofing system, the ceiling and all its components (i.e. drywall, light fixtures and hanging panels) had to be suspended on a spring like system, allowing only a 2.54cm tolerance in vibration. Structural columns and exterior walls were cut 7.62cm at the base, also for the purpose of allowing for vibration tolerance.

这家夜总会的老板有机会重新定义加拿大尼亚加拉瀑布的夜景，并可使这家夜总会成为瀑布景观娱乐度假村零售和娱乐中心的一部分，对此，他十分兴奋。为了使游客保持一整天的兴奋和活跃的状态，就需要在加拿大尼亚加拉大瀑布的旅游城市中创建新的夜间娱乐场所。城市对瀑布景观娱乐度假村项目的投资为建造一个新的拉斯维加斯风格的娱乐场所提供了机会。度假村的管理公司认为，一个新的夜总会将有助于提升瀑布景观娱乐度假村的欢乐氛围；因此，这个面积为 1115 平方米的提供全方位服务的娱乐场所预期会容量约 800 人。除了提供一流的服务外，这个娱乐场所还有全套的多媒体展示设施，可以承办公司活动、时尚节目和促销活动。

在功能方面，设计力图创建一个多功能娱乐场所；在创意方面，则希望营造一种神秘且具有异国情调的环境，以便吸引来自世界各地的夜店爱好者。最终，设计希望打造一个让人忘却外界喧嚣的娱乐圣地，让声音、灯光、人和装饰和谐相融，来换取一次激动人心的超越时间和空间的体验。在设计过程中遇到的最大挑战是隔音方面的问题。建筑的开发商没有考虑到夜总会会被放置在大楼的最底层，因此，需要特别的工序设计，来保证夜总会营业时不会干扰到邻近的商家及酒店的客人。为了加入一个适当的隔音系统，天花板和所有的组件（如石膏板、照明固定器、吊板）必须被悬挂在一个像弹簧的装置上，并且只允许 2.54 厘米的振动公差。同样因为振动公差的关系，结构柱和外墙在底部被缩短了 7.62 厘米。

EXIT

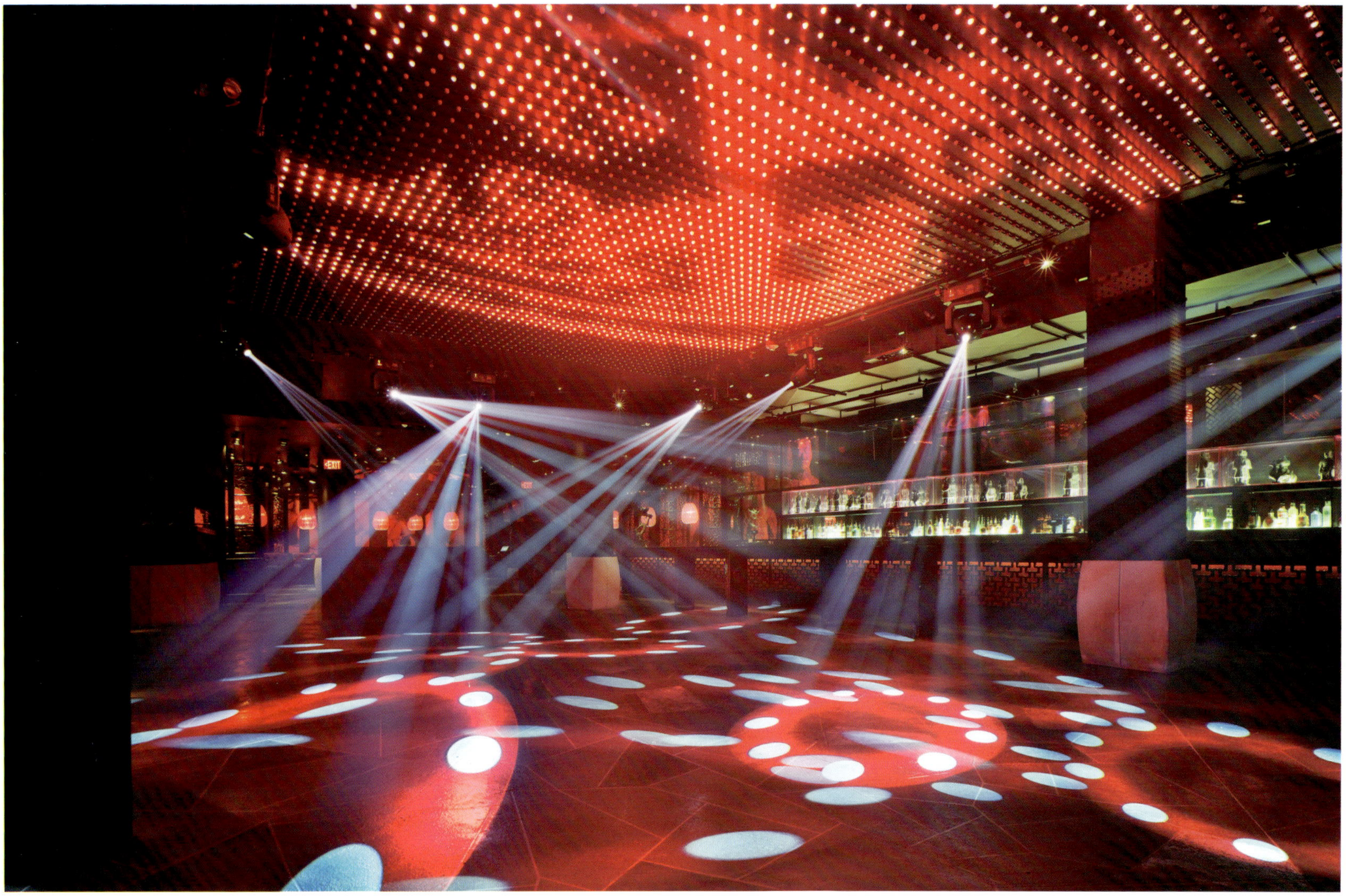

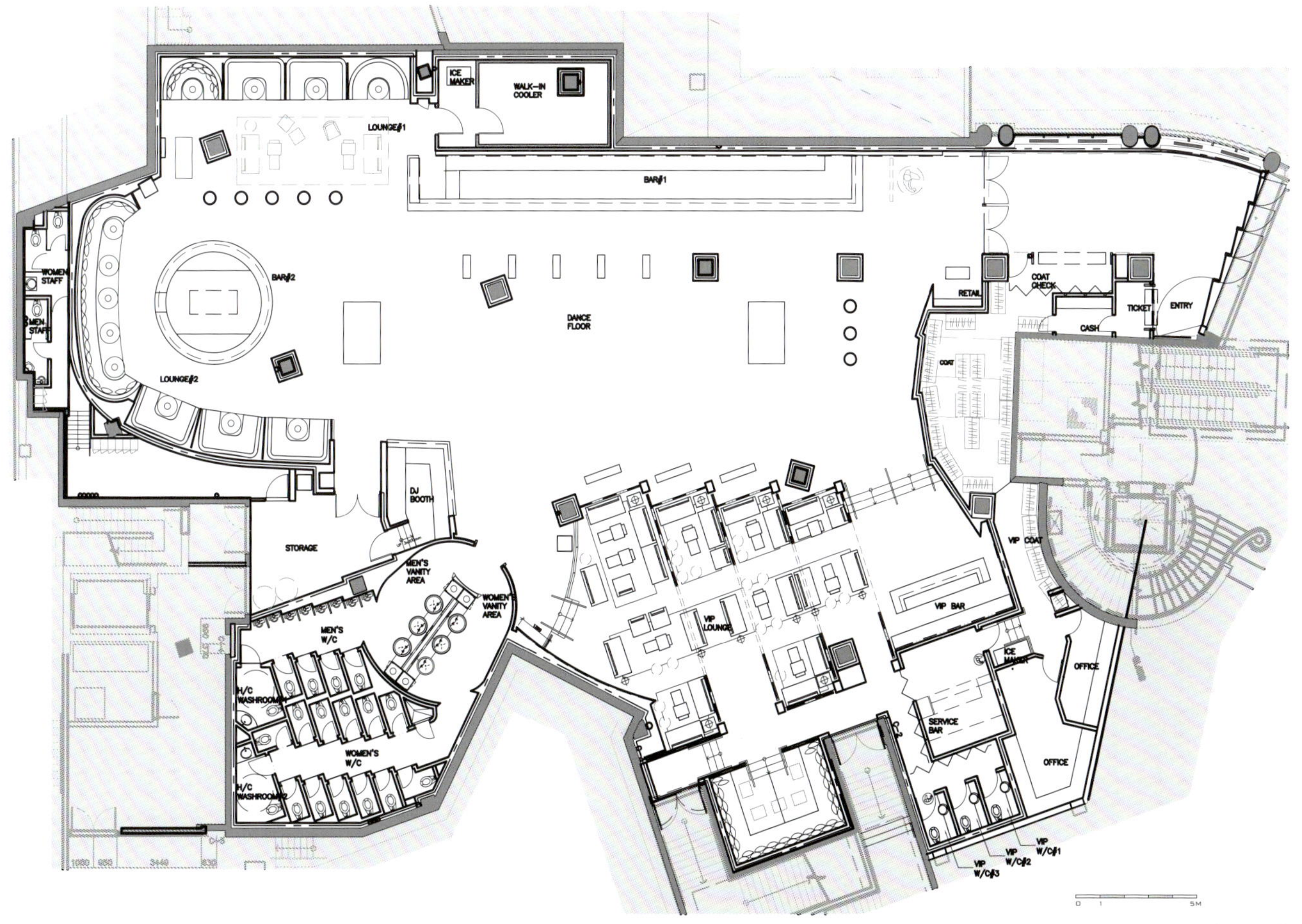

ICE MAKER
WALK-IN COOLER
LOUNGE#1
BAR#1
BAR#2
WOMEN STAFF
MEN STAFF
DANCE FLOOR
LOUNGE#2
RETAIL
COAT CHECK
TICKET
ENTRY
CASH
COAT
DJ BOOTH
STORAGE
MEN'S VANITY AREA
WOMEN'S VANITY AREA
MEN'S W/C
H/C WASHROOM#1
WOMEN'S W/C
H/C WASHROOM#2
VIP LOUNGE
VIP BAR
VIP COAT
ICE MAKER
OFFICE
SERVICE BAR
OFFICE
VIP W/C#1
VIP W/C#2
VIP W/C#3

Yusaku Kaneshiro+Zokei-Syudan Co.,Ltd

Hotel Club – SUIREN

Designer
Yusaku Kaneshiro, Hiromi Sato

Location
Nagoya, Aichi, Japan

Area
126 m²

Photographer
Shinichiro Sugai

Glamorous colors and coffered ceiling in the pattern of flower construct modern Japanese space. Impressive high heel arts and chandeliers are arranged in the space.

Display of glass case and small feminine articles in it image "a treasure box."

Though we aimed to create a space where women look more beautiful, also could realize the one that is far a part from daily life with using graceful curves.

迷人的色彩和带有花型的方格天花板共同建构了现代的日本空间。在这个空间内还装饰着令人印象深刻的高跟鞋艺术品和吊灯。

玻璃橱的陈设和小女人的饰品象征着这个空间是一个“宝箱”。

尽管我们的目标是要创造一个让女人看起来更漂亮的空间，但实际上，通过使用优美的曲线，我们还是使该空间远离了生活。

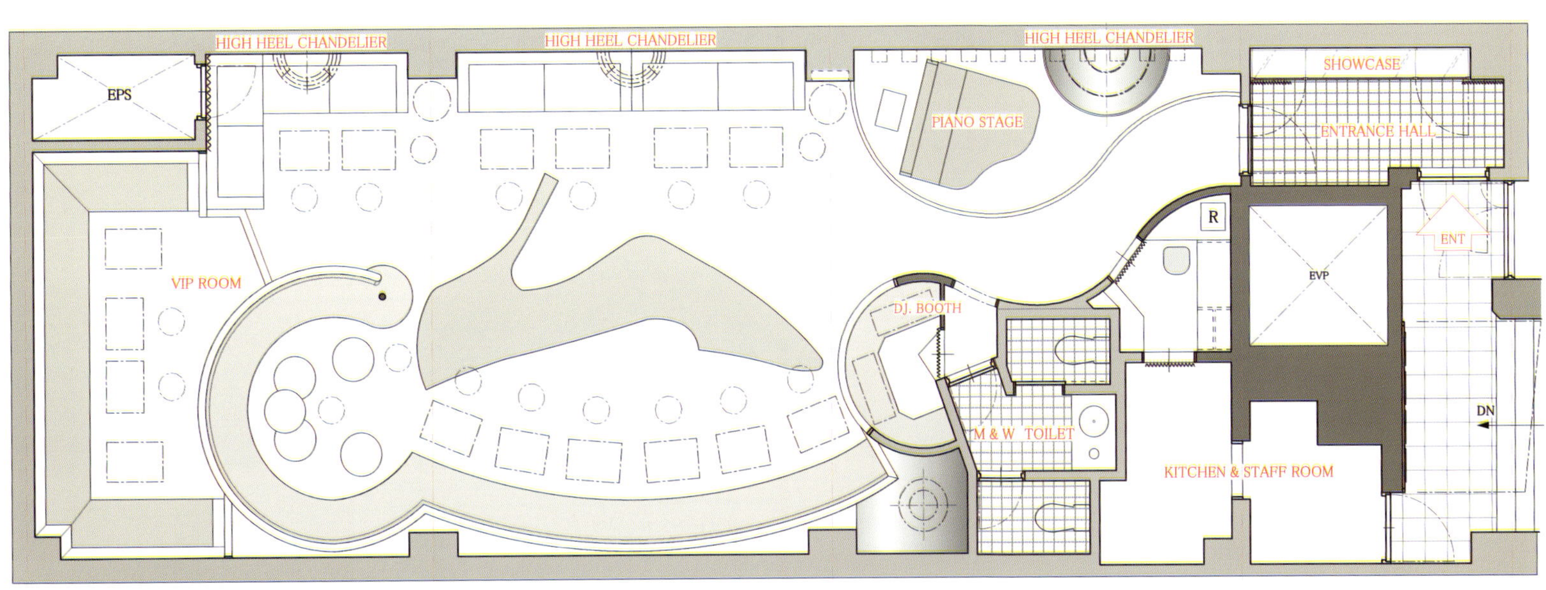
HIGH HEEL CHANDELIER
HIGH HEEL CHANDELIER
HIGH HEEL CHANDELIER
SHOWCASE
EPS
PIANO STAGE
ENTRANCE HALL
R
ENT
VIP ROOM
EVP
DJ. BOOTH
DN
M & W TOILET
KITCHEN & STAFF ROOM

Elia Felices Interiorismo

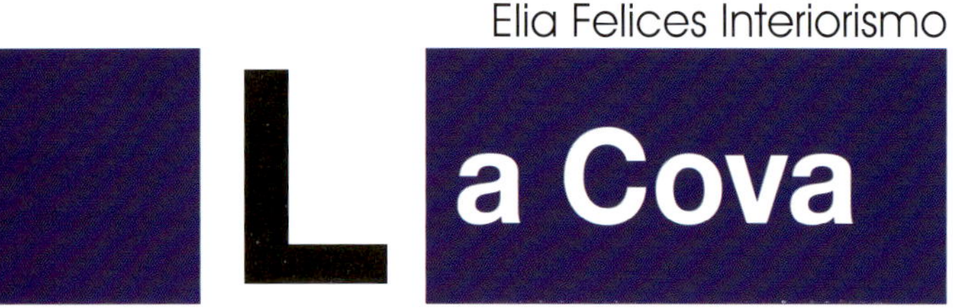

Designer
Elia Felices

Location
Mataró, Barcelona

Area
500 m^2

La Cova. Dance Theatre, located in Mataró, near Barcelona, is a nightclub with a total area of 500 m^2. It is equipped with variable lighting, furniture designed exclusively by the studio, and forms and colors that replicate ice caverns. A landscape of this type in the middle of the city causes a great impact.

The interior is a flowing space with different areas, most of which are open and easy to move around in, creating circuits in which people can interact. To emphasize this ease of movement within the public areas, the frontage is made of transparent glass. The project is laid out as a long space on a single level which allows for a rational and intense distribution of the clientele. There are a number of different areas: The cloakroom and ticket office are on one side of the entrance. A curved wall with an elegant blue velvet curtain leads us through a door similar to that of the facade and onto the dance floor where two bars with long counters run along either side, and the entrance to the toilets is also here.

The clear white ticket office is decorated with a floral mural in underwater blue shades. An elegant doorway opens onto another larger mural similar to the first. The bars feature perspex ovals backlit by RGB leads where fish can be seen swimming and hiding themselves among the sea plants where they change color. There are also large MDF rosettes painted white. The same can be seen in the DJ cabin and cloakroom. This scenery is continued through the backlit bottle racks, where the bottles form part of a play of lights, transparencies and colors associated with a glaciated landscape.

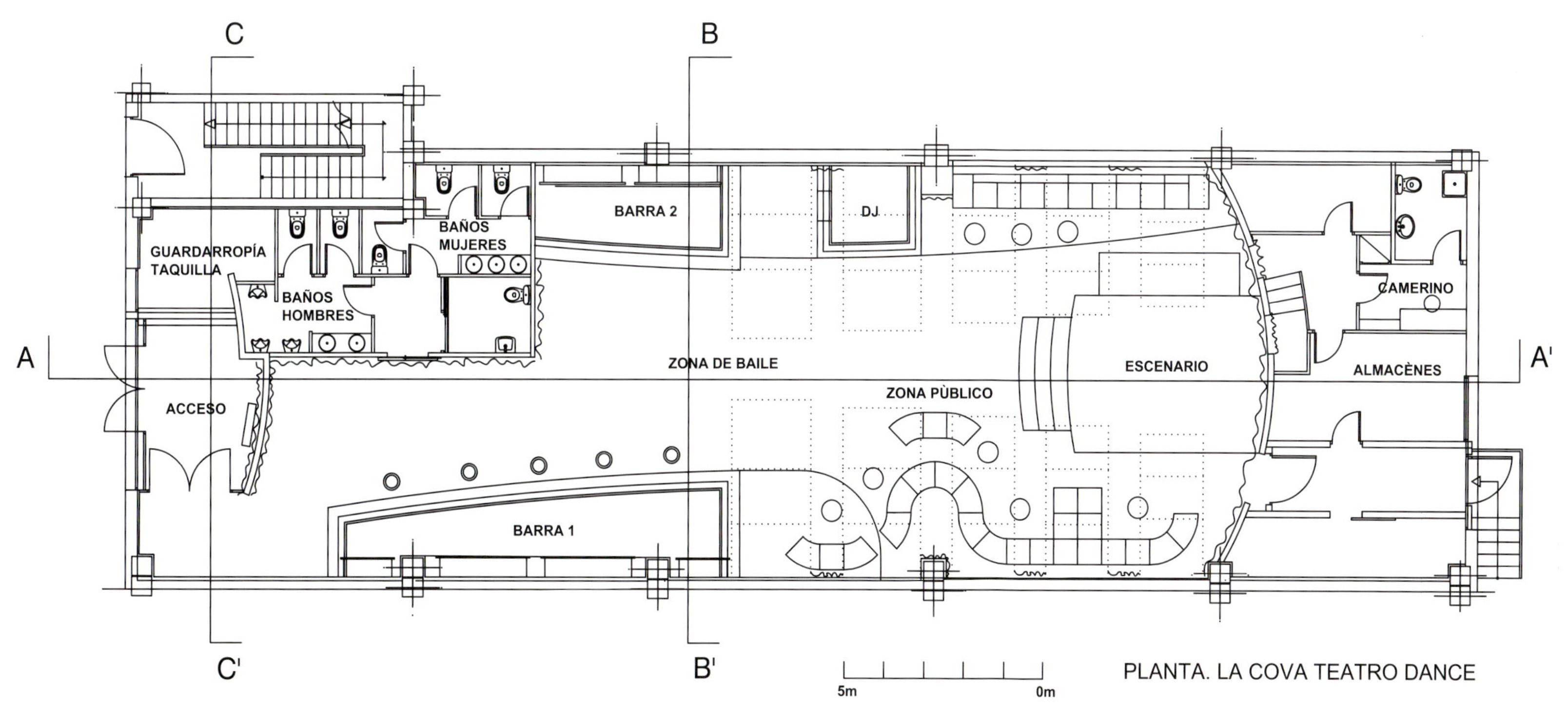

PLANTA. LA COVA TEATRO DANCE

La Cova 舞厅位于巴塞罗那附近的马塔罗，是一个总面积为500 平方米的夜总会。夜总会配备了可变光，家具完全由工作室特别设计，造型和色彩方面的设计则模仿了冰洞。此种类型的景观设计在城市中心引起了极大的反响。

室内是一个包含不同区域的流通空间，大部分区域都是开放并且易于走动的，环形路的设计方便了人与人之间的互动。为了强调公共区域流通的特性，舞厅的正面是由透明玻璃构成的。该项目为水平方向上的长条形空间布局，此种布局方式能够有效合理地分散客流。夜总会内分为许多不同的区域：靠入口处为衣帽间和售票处。一堵弯曲的墙和一条优雅的蓝色丝绒窗帘引导客人穿过一扇与外墙十分相似的门，到达舞池，两个酒吧的长柜台分布在舞池两侧，厕所的入口也在这里。

纯白色的售票处装饰着一幅水底蓝色渐变色调的花样壁画。一个优雅的门廊对面是另一幅更大的壁画，壁画内容与前一幅十分相似。酒吧的亮点是带有红绿蓝背光的弧形玻璃，透过玻璃，宾客可以看到里面游来游去的鱼儿，当游到海草中时，随着灯光的变化，鱼儿的颜色也会随之改变。室内还装饰有由中密度纤维板制成的大花环，花环被涂成白色，被应用在 DJ 台和衣帽间等处。此种装饰手法在背光式瓶架的设计上得以进一步体现，此处，酒瓶也成了光艺术的一部分，整体设计的透明感和多变的色彩让人很容易联想到冰川景观。

Sugarcane

Designer
Lionel Ohayon & Siobhan Barry

Photographer
Francis + Francis

Sugarcane, the boutique nightclub concept of Sushi Samba, is separated from the restaurant by a long corridor that acts as a transition space between two types of performances where guests enjoy a respite from the high-energy, carnival inspired restaurant. At the end of the corridor, Sugarcane, a Favela-inspired late-night space has a muted palette to create a sultry atmosphere that pays homage to the iconography of Sugarcane's key DNA. A stage, dance floor and DJ booth are highlighted by materials that have a raw feel – gold and chocolate brown leather, bamboo, rustic wood, smoked and frosted acrylic, resin and paper. A custom-designed, "sugarcane" ceiling is a combined architectural and lighting element. Moving lights wind through tubes and rods that begin as a ceiling element and become performance poles. The VIP area is styled to include plush seating, mood lighting, Third World relics and Brazilian Favela street finds. The public space of the club flows to the restrooms, creating vistas into and out of the rooms and playing with the concept of privacy as a performance space. Bathrooms "pods" are scaled to recall oversized sugarcane and are colored with graffiti by Brazilian artist Felipe Yung "Flip".

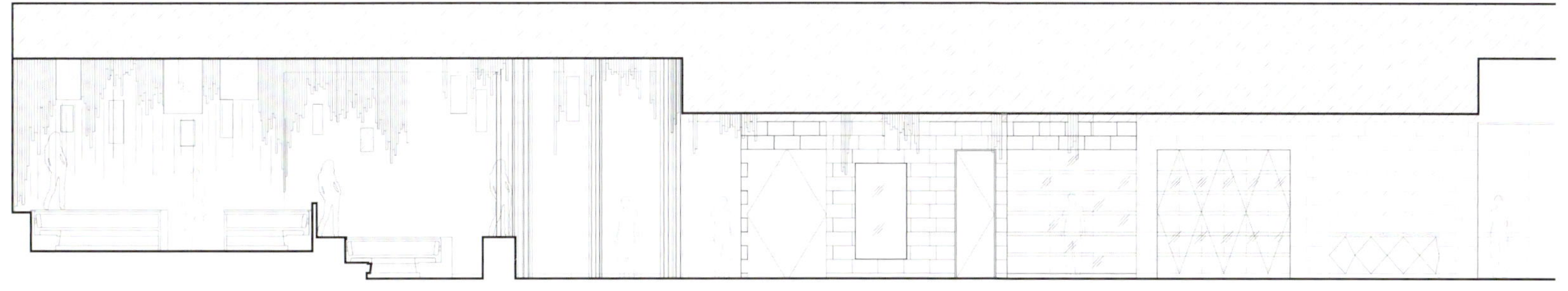

延续Sushi Samba的精品夜总会理念，Sugarcane与餐厅以长廊相隔。这条长廊仿佛是两种风格的过渡地带，也可以说是狂欢味儿十足的餐厅的一个缓冲地带。长廊尽头是受贫民区风格影响的午夜场Sugarcane。暗色基调延续了Sugarcane惯有的标志性迷人氛围。舞台、舞池、DJ台的材质是其设计亮点，包括皮革、竹子、原木、磨砂合成亚克力、树脂和纸张，以金色和棕色为主要颜色，给人一种粗犷的感觉。定制的“甜蔗”天花板是建筑与灯光的完美融合。摇头灯在长短不一的灯管或灯柱间蜿蜒排列，灯柱起初只是天花板的装饰，后来则演变成了表演柱。贵宾区混合了豪华坐椅、感应灯、第三世界遗迹和巴西贫民街头风格。夜总会的公共区域延伸至休息室，营造出进出休息室的远景，并将私密空间当成表演空间一般精心设计。厕所厕位被设计成了大号甜蔗的造型，并且由巴西艺术家菲利浦•杨创作的“空翻”涂鸦所装饰。

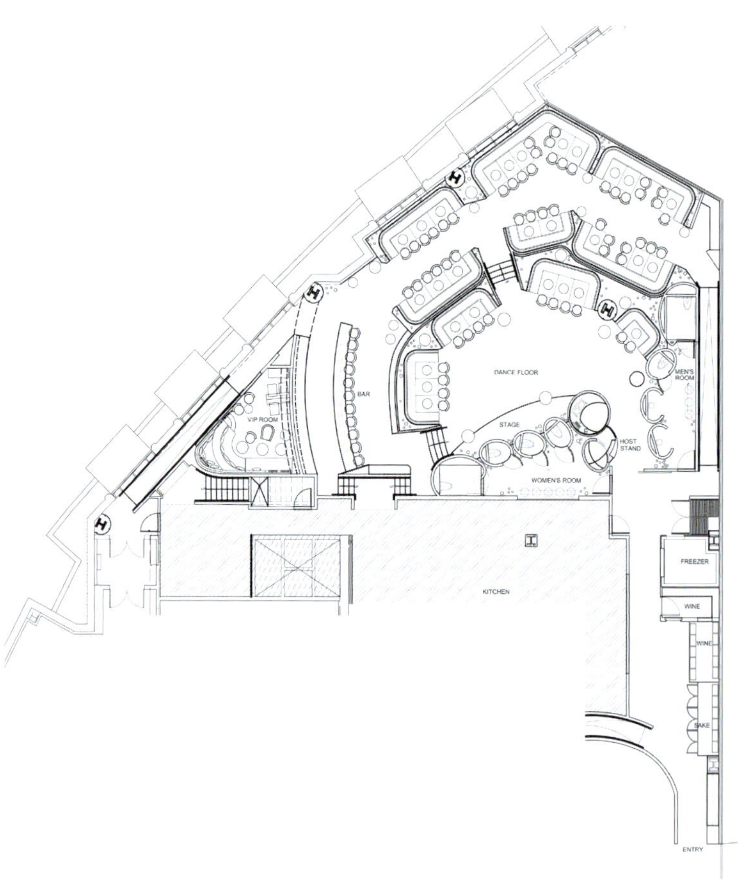
DANCE FLOOR
MEN'S ROOM
BAR
VIP ROOM
STAGE
HOST STAND
WOMEN'S ROOM
FREEZER
KITCHEN
WINE
WINE
SAKE
ENTRY

Bluarch Architecture + Interior + Lighting

Designer
Antonio Di Oronzo

Location
New York, USA

Photographer
ADO

Greenhouse is built from recycled or recyclable materials. Greenhouse is the first nightclub in the nation to receive certification via LEED_CI by the United States Green Buildings Council.

The designer decided to stay away from re-creating a Greenhouse, and opted to transpose the notion of landscape to an interior space. The design concept was to convey the dynamic richness of nature as a living system. The walls connect to the ceiling via a series of laser-cut ribs creating a shelter within the space. The ribs are lined with series of 0.15 meters round panels organized in a self-similar and recursive pattern generated through a fractal algorithm. One third of these panels are upholstered with eco-friendly vinyl, one third of them are clad in sustainable boxwood. The rest is lacquered and each disk houses an LED light point of 0.8 watts. The light points, in total 2,500, are all connected to custom designed software which allows for maximum lighting flexibility. These LED pixels can describe effects which derive from music beats or video signals.

The ceiling is an organic formation of 40-millimeter crystals representing a body of water about to project onto the ground. The crystals are seemingly passively appended, but slightly vibrate in the music and vividly respond to the green lasers and the LED.

The tables are a quirky presence in the overall scheme. They are tempered glass boxes containing boxwood covered wireframes shaped as different animals resting on a carpet of eco-friendly, artificial grass.

The bar is a scale model of a gently sloping landscape punctured with miniature trees and scale models of house designed in the past. The model is seemingly a straight cut through the crust of the earth.

The main challenges in a LEED_CI certification process for commercial interiors are retrofitting existing systems to meet strict efficiency requirements. The designer set out to achieve the same aesthetic freedom and quality while making the intervention ecologically acceptable.

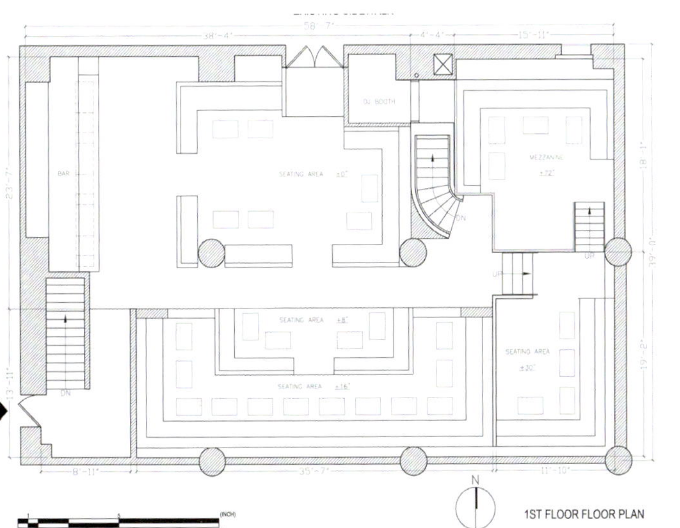
1ST FLOOR FLOOR PLAN

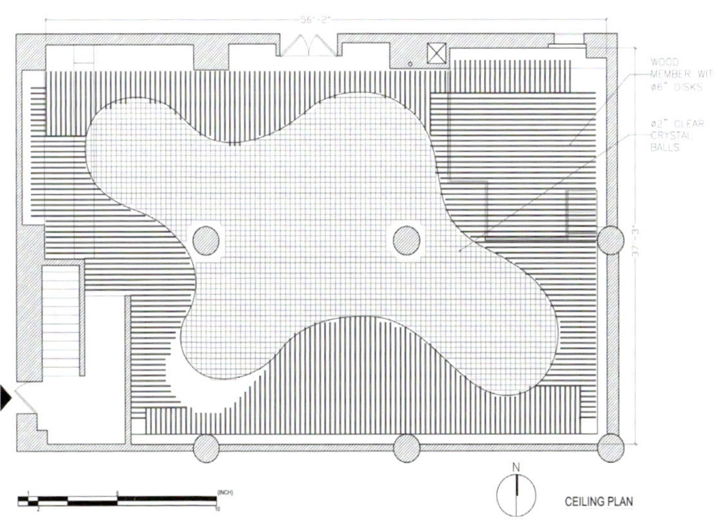
CEILING PLAN

Greenhouse在建筑中全部使用再循环材料或可再生材料，是纽约第一个获得由美国绿色建筑委员会颁发的LEED_CI(绿色建筑评估体系商业室内装修)认证的建筑。

设计师没有原样照搬一个Greenhouse，而是选择把景观设计的理念用于室内空间设计中。设计理念是把生机盎然、丰富多彩的大自然生动地展现出来。一系列激光切割的拱肋把墙壁和天花板连接起来，在空间中构成一个遮蔽棚，拱肋中内衬一系列直径为0.15米的圆形嵌板，按照分形算法生成的自相似循环图案排列。三分之一的嵌板由无污染的乙烯基材料包覆，还有三分之一的嵌板用可回收的黄杨木覆盖。其余嵌板则被刷上漆，每块板子上都安装了一个0.8瓦的LED灯，总共安装了2500个LED灯，并都连接到专用的软件上，以提供最灵活的照明控制，这些LED灯还能根据音乐节拍或者视频信号变换照明效果。

天花板用直径为40毫米的水晶颗粒组成了一个有机的整体，象征着一池清水，波光投射到地面上。水晶颗粒看上去似乎是简单地连接在一起，但是却可以随着音乐微微颤动，并能灵敏地反射绿色激光束和LED灯光。

整个方案中，桌子的设计比较标新立异。桌子被设计成一个个的钢化玻璃箱子，里面是一只只动物的外形轮廓，轮廓外包覆着黄杨木，下面铺设着一层由环保材料制成的人造草坪。

整个吧台是一个缓坡景观设计的缩小模型，上面点缀着树木模型和一些设计师以前设计的房屋模型，这个模型就好像是从地球表面直接切下来的一样。

本案商业室内装修申请LEED_CI认证的最大的挑战在于改进现有的系统，使之满足严格的效能要求，为此，设计师努力在不影响美学自由和美学质量的同时，达成生态环保的目标。

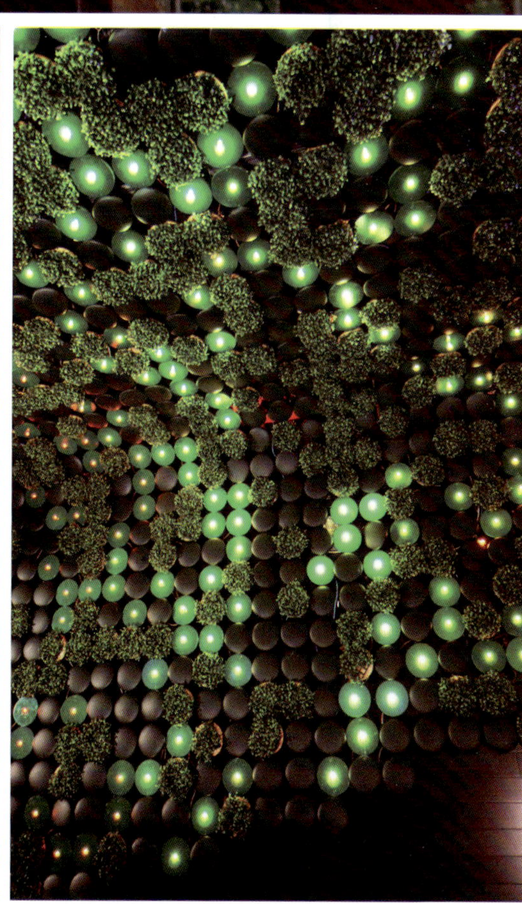

Archinexus

Room 18 + 18 Lover

Designer
KAN, TAI LAI

Location
Taiwan, China

Room18

Within a theatrical environment, double-bar form is introduced, which shapes a multi-oriented and multi-leveled dance pool among the bars and the DJ booth and redefines the concept of a bar with the notions of "boy bar" and "girl bar". Here, you are audience and star at the same time, for the stage is everywhere you go and you can show your passion at anytime you want.

18Lover

A series of vintage-feel ribbon and linellae are laser cut on multi-layered transparent acrylic sheets which form translucent screens partitioning the seats, layer upon layer, blending and mixing, thus a neo English digital garden is shaped with illusive reflections and flowing lights and shadows.

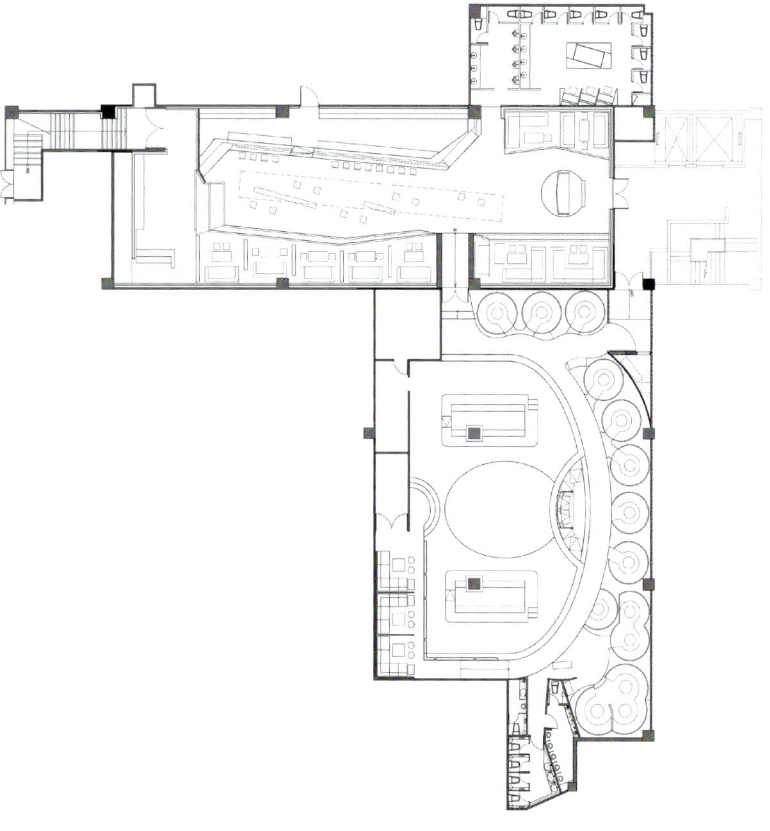

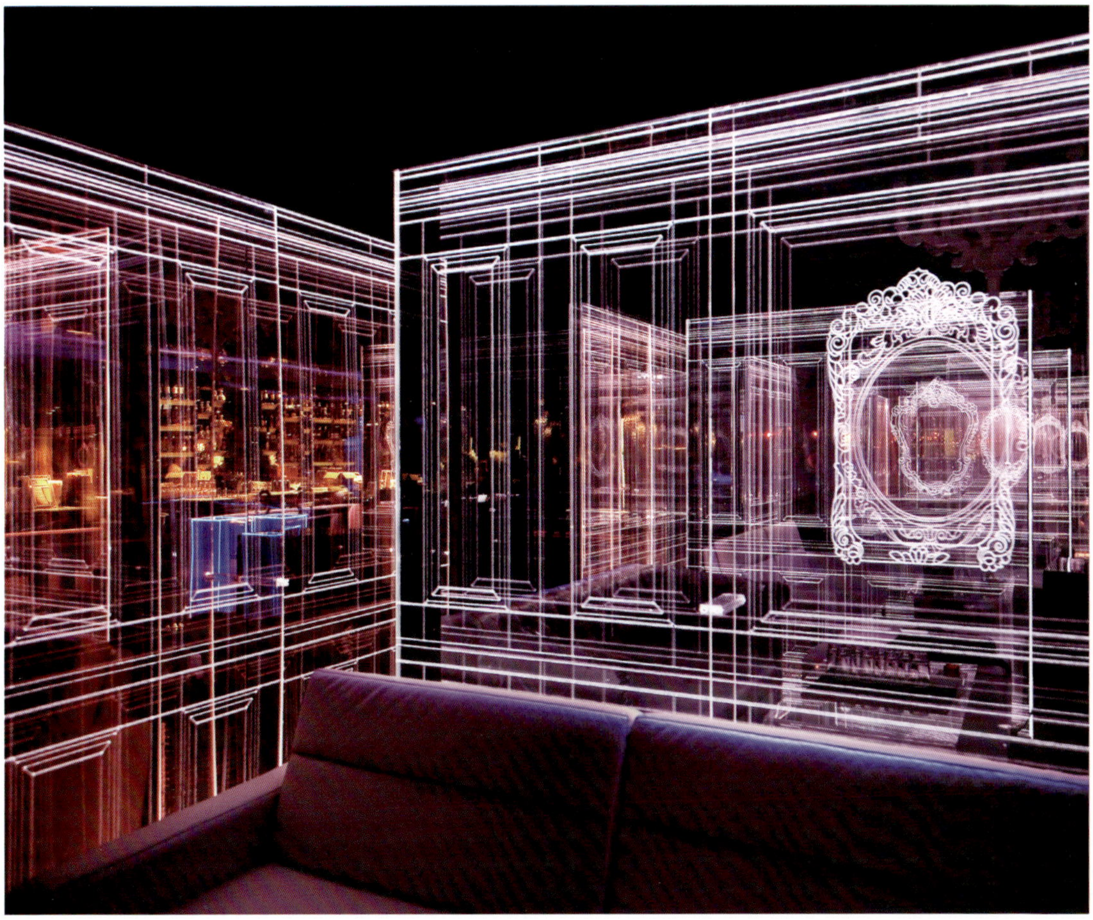

Room18

在“剧场式”的空间环境架构中，透过一组双吧台的形式置入，使其和DJ台之间形成了一个多面向、多层次的舞池，同时，亦衍生成新的吧台概念：“男生吧”及“女生吧”。在此，客人既是观众又是明星，因为随处皆是舞台，随时皆可展现奔放的自我。

18Lover

设计将一系列复古风格的饰带和线条雷射雕刻于多层次的透明亚克力板上，使其成为座席间半透明的隔屏，层层交迭融合，进而形塑出一处虚实相间、光影变幻的新英式数字花园。

Parolio & Euphoria Lab

Pacha Madrid

Designer
Parolio

Location
Madrid, Spain

Area
1,475 m^2

Photographer
Ricardo Labougle

Pacha Madrid by Trapote Group occupies since 1980 one of the only art deco buildings in the city of Madrid, through the years the decoration and design of the interior didn't play up this amazing location.

The new image designed by Parolio experiments with shapes and materials used in the early 1930s with a renovated and fresh take, streamline and dramatic design sense. The outcome is one of Spain´s most incredible venues.

The project is divided in 4 specific areas:

The stage and VIP area

This multi-use space is used as a stage or VIP area, Parolio created a design of modular furniture that could be mounted or dismounted easily, and create different shapes of sofas for different sessions or parties.

Main floor

A 10 meter LED light and glass installation covers the wall of the main floor.

The main bar is a semi circular shape with a motored / rotating centered bottle rack.

Two additional bars have mini screens for video and advertisement, and geometric bottle racks made of stainless steel and mirrored glass.

Voyeur balcony

Another VIP space is the man balcony that overlooks the main floor, the walls are covered by padded geometric paneling and mirror.

The DJ table is suspended from the center of the balcony with a glass and mirror glass structure.

Lobby

Sculptural backlit mirror mural, geometric deco lamps made of glass and mirror glass and the floor is completed by a black and purple geometric design tiled carpet.

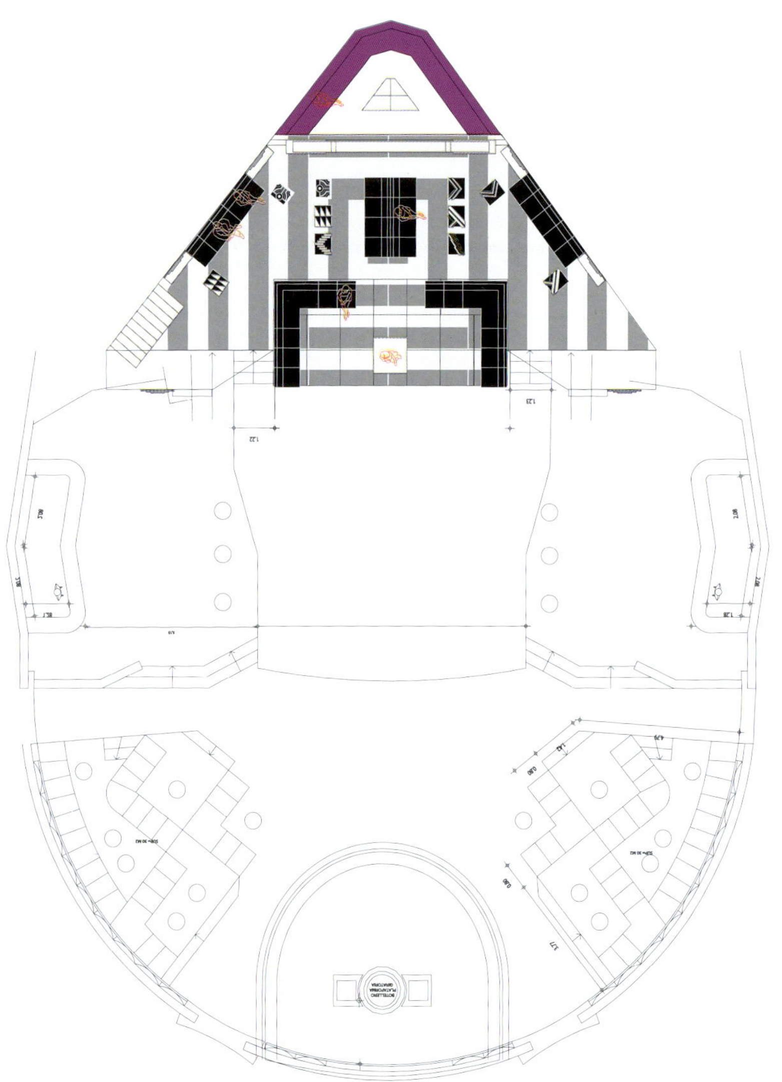

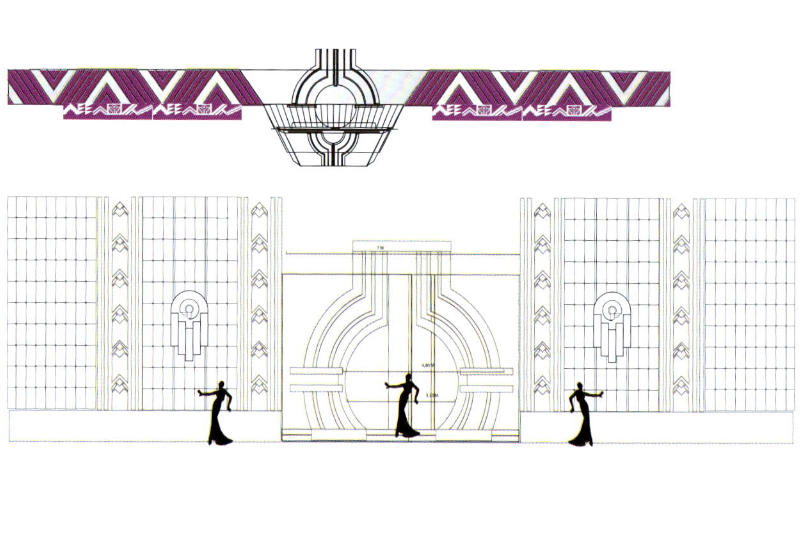

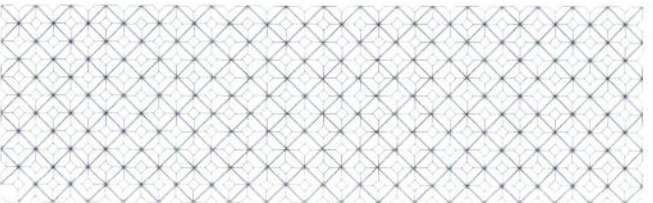

从1980年至今，Trapote集团旗下的Pacha Madrid就是马德里市唯一的具有装饰艺术风格的建筑，多年过去，建筑的室内装修已经不再与其所处的绝佳位置相匹配了。

设计师Parolio尝试使用在20世纪30年代早期常用的装饰形状和材料，通过翻新旧材料营造新表情。采用直线型和戏剧化的设计手法，赋予了建筑全新的形象，最终使这里成为了西班牙最不可思议的场所之一。

该项目可细分为4个区域：

舞台和贵宾区

一个多用途的空间被用作舞台或贵宾区，Parolio设计了一种模块化的家具，易于安装和拆卸，并可根据会议或聚会形式的不同而组合出形状不同的沙发。

主层

一个十米长的LED照明灯和玻璃装置覆盖着主层的墙壁。

主吧台为半圆形，内有一个动力/旋转中心的瓶架。

两个附属吧台有播放视频和广告的迷你屏幕，以及由不锈钢和镜面玻璃制成的几何形瓶架。

观看包厢

另一个贵宾空间是观看包厢，可俯瞰主层，墙上覆盖着几何镶板和镜子。

DJ台悬挂在包厢的中心，为玻璃和镜面玻璃结构。

大厅

大厅内有带有背光的雕刻镜面壁画，由玻璃和镜面玻璃制成的几何状装饰灯具，地面则铺有黑色和紫色的几何形的地毯。

TAI Architecture & Design Inc.

Story Nightclub

Designer
Francois Frossard

Location
Miami Beach, USA

Photographer
Simon Hare

Coined "the brand new home for South Beach's party monsters", Story Nightclub harbors an immensely large two storey space that can easily accommodate over 2000 people, accompanied with 56 VIP tables and 5 accessible bars. The design of the space is set in dark tones to emphasize the state of the art lighting system, one of the most complex and unique in the nation. The main dance floor is surrounded by six twisted spiraling columns wrapped in dark brown croc skin vinyl and accentuated with pixel mapping LED strip lights. Suspended above the dance floor is a star shaped lighting rig with 3,264 ping pong sized LED pixel balls, each one individually controlled by the lighting and video jock. All the energy is focused toward the oversized DJ booth which is flanked by 5 huge LED video walls and its own private VIP seating area. The mezzanine floor is wrapped on the perimeter with custom mirror LED wall panels which are synched with the rest of the show lighting. All combined, the lighting show creates a unique atmosphere that leaves patrons something to "wow" about the next day. Also adding to the night's experience is a custom designed Infinite sound system which can pound your soul from anywhere within the club.

Though Story's layout creates the sense that the nightclub is focused on the bling and VIP bottle service, its open plan has plenty of area for clubbers to dance and mingle and enough bars to avoid any queuing in line for a drink. The newly revamped space exudes musical madness combined with chic elegance welcoming patrons to have a memorable nightlife experience and have their own individual "Story" unfold.

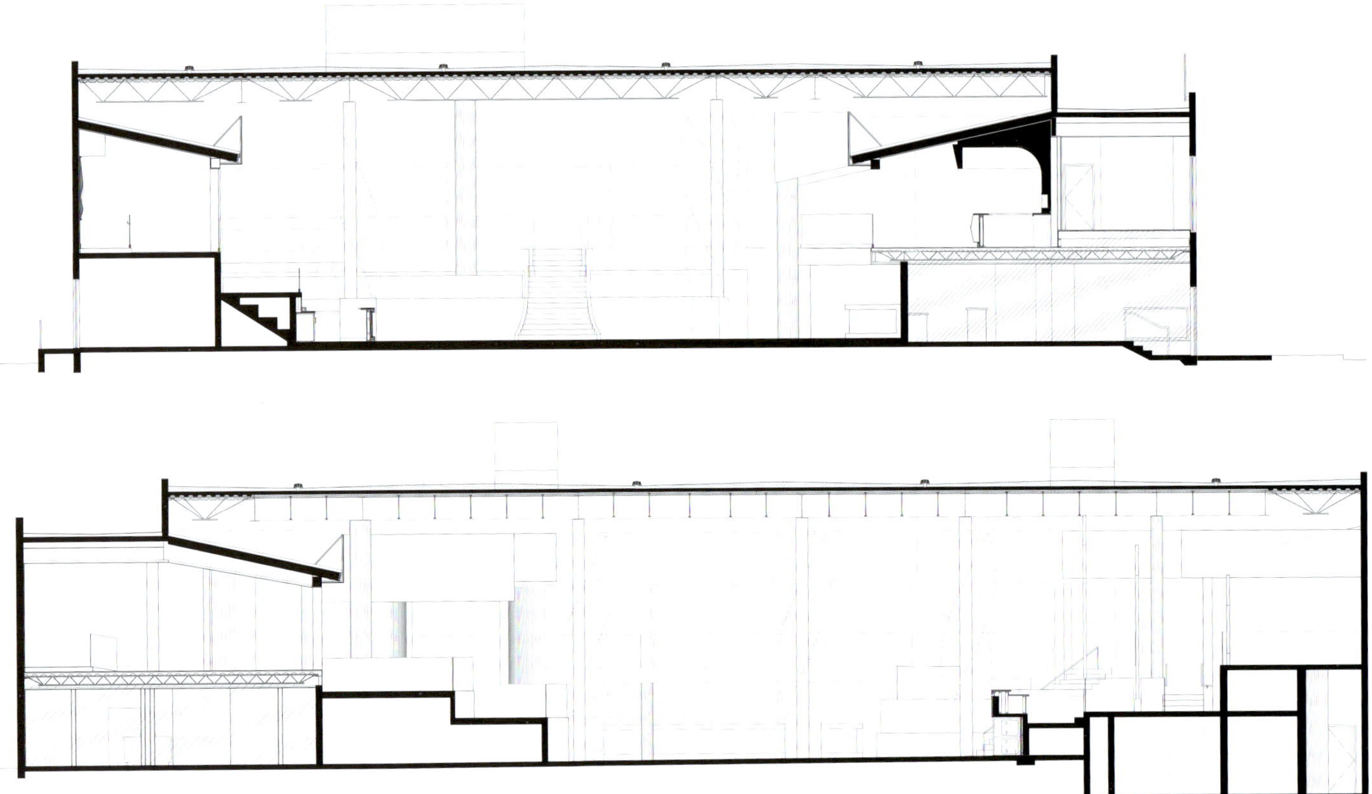

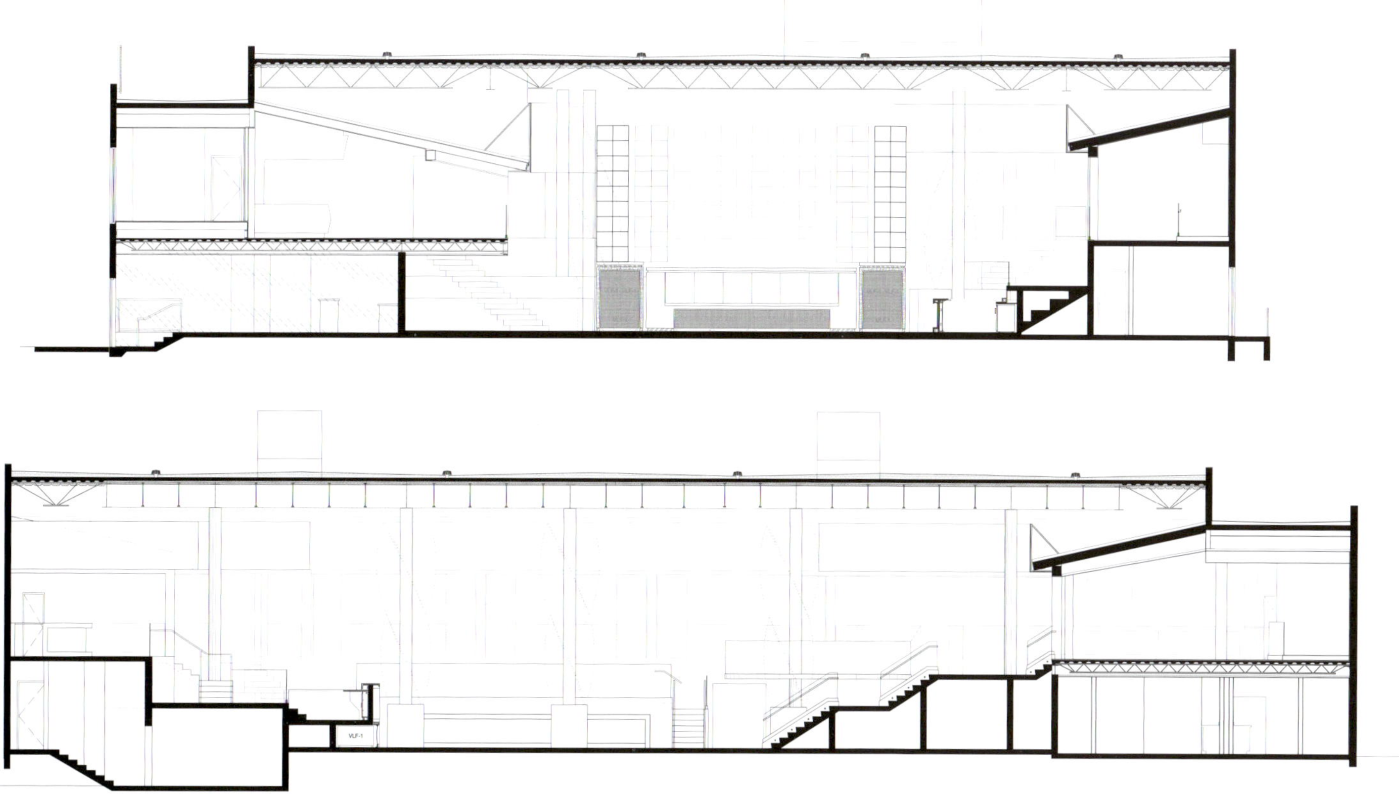

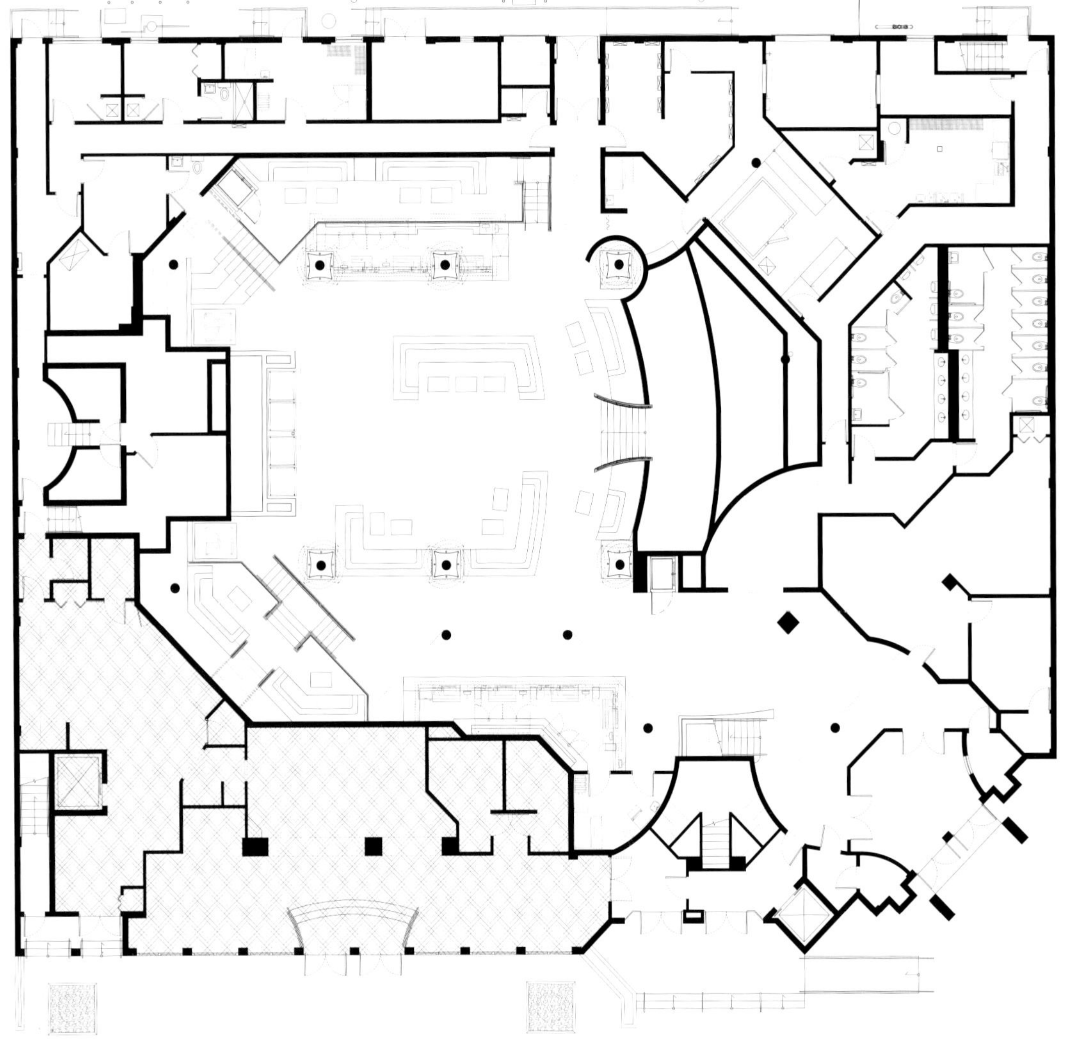

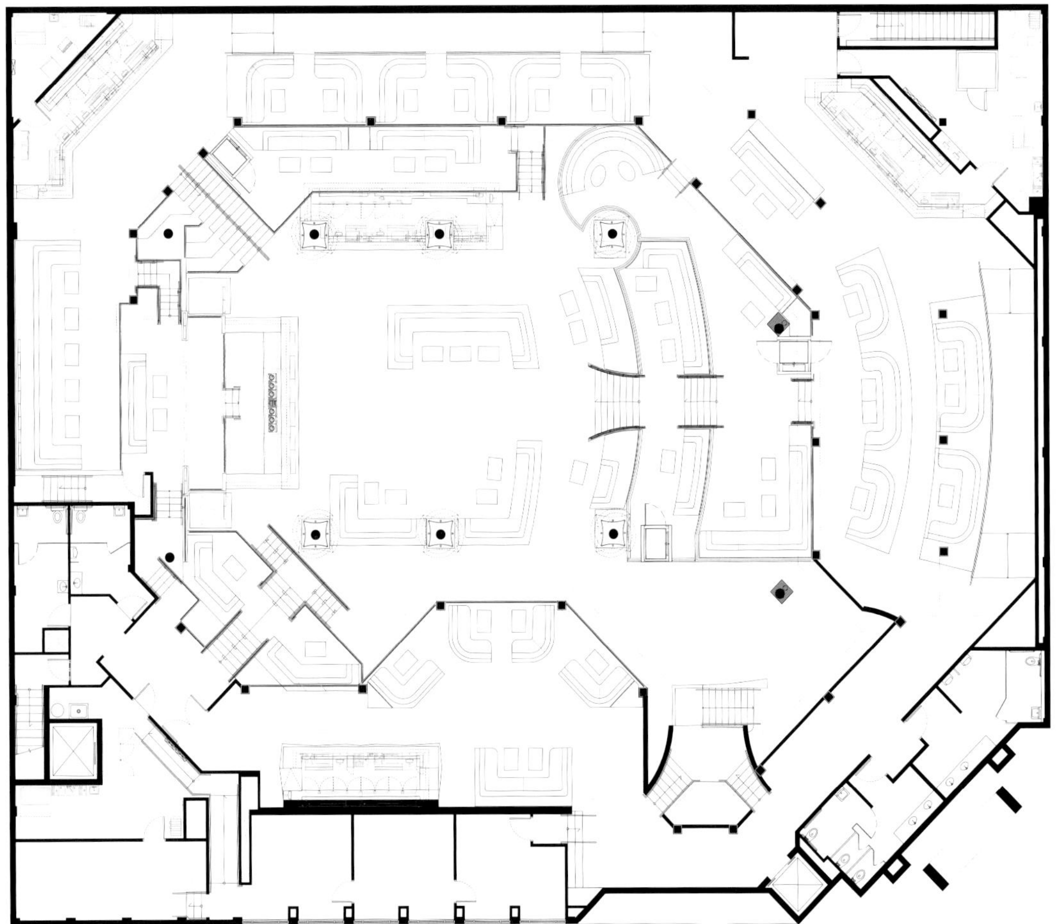

Story夜总会铸就了“南海滩社交达人们的新家”，该项目共两层，拥有超大空间，可以轻松容纳2000多人，同时还配置了56张贵宾桌和5个可利用的吧台。室内的空间设计以暗色调为主背景，突出了先进的照明系统——美国最复杂和最独特的照明系统之一。主舞池周围有六个旋转的螺旋列柱，柱子外包裹着黑棕色鳄鱼皮乙烯基，并点缀着像素映射LED灯带。舞池上方悬吊着星形照明设备，由3264个乒乓球大小的LED像素球组成，分别由照明和视频人员控制着。所有的能量都集中朝向超大的DJ台，DJ台周围是5个巨大的LED视频墙以及私人贵宾座位区。阁楼的外围包裹着定制的镜面LED护墙板，与其他灯光照明同步。将所有元素组合在一起，声光色电俱全的表演创造了一种独特的气氛，留给顾客的将是第二天的惊喜连连。同时，给夜生活锦上添花的还有一套定制的音响系统，无论你在夜总会的哪个角落，都可以感受到震撼心灵的音乐冲击。

虽然Story夜总会的布局会给人一种灯光闪烁和贵宾式服务的感觉，但它开放式的规划却提供了足够多的空间供会员跳舞和互动，并且还设有足够多的吧台，避免了为饮品排队的问题。新装修的空间散发着音乐的魔力与时尚的优雅，欢迎顾客光顾，享受难忘的夜生活，开启属于自己的故事。

Concrete Architectural Associates

Supperclub San Francisco

Designer
Rob Wagemans, Onne Walsmit

Location
California, U.S.A.

The salle neige of the Supperclub San Fransisco is built like an arena. The bar and kitchen are positioned in the middle of the room, so the kitchen staff can become the DJ of the night.

The disabled toilets are the most spectaculair thing about this Supperclub, this is our reaction on the idiot regulation in California regarding disabled people. There big, luxurious and pink Handcuffs and whips are attached on the walls. The pink skaileather cushions are creating the best atmosphere and surrounding for unexpected and exiting meetings.

What will be next in the forthcoming venues of London, Singapore, Istanbul and maybe Singapore.

Supperclub is one of a kind and therefore very successful.

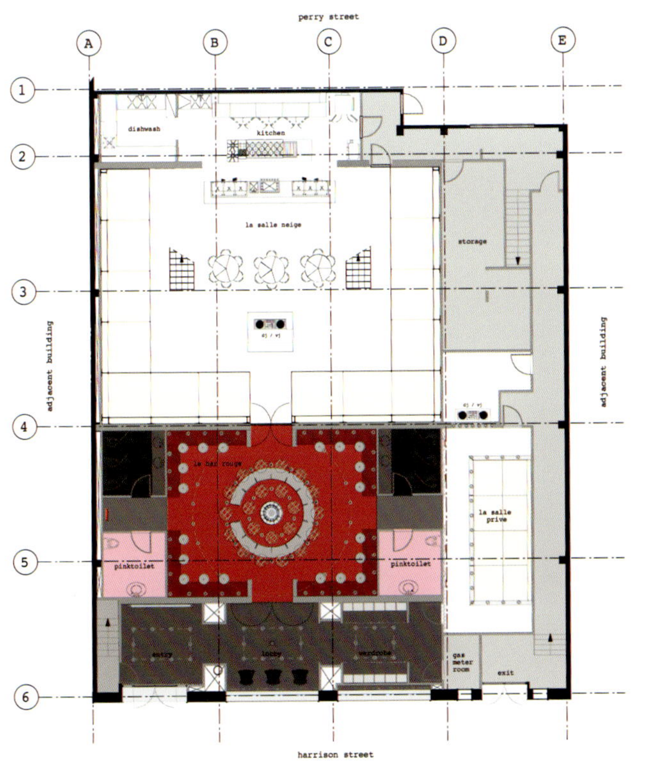

first floor

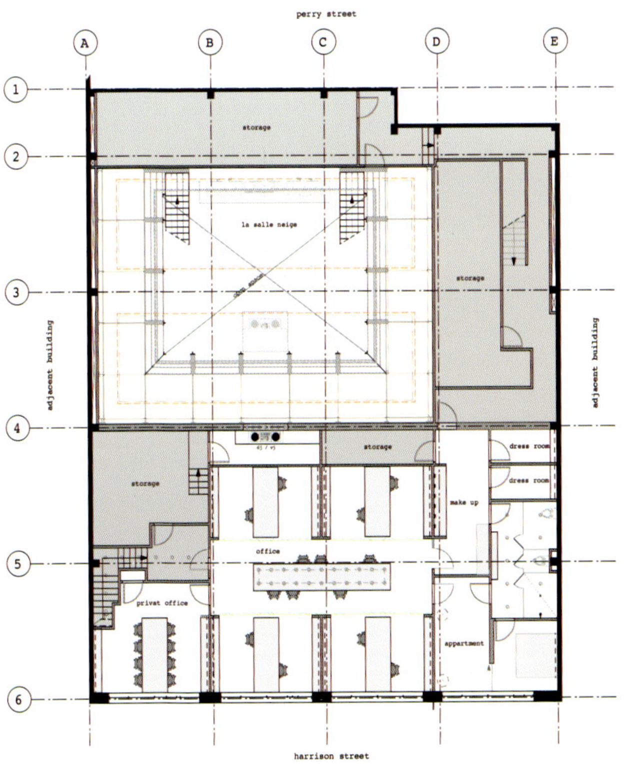

second floor

EXIT
EXIT

EXIT
EXIT

旧金山夜总会大厅的外形像一个舞台。酒吧和厨房位于夜总会的中央，因而厨房的工作人员在夜晚可以成为 DJ。

残障厕所是旧金山夜总会最特别的地方，该设计也是设计师对加州关于残疾人的白痴规定的回应。墙上挂着华丽的粉红色的大手铐和鞭子。粉色的皮革垫子为意想不到的且令人兴奋的会面营造了最好的氛围和环境。

下一个 Supperclub 会建在哪里？伦敦、新加坡，还是伊斯坦布尔？也许是新加坡。旧金山夜总会是独一无二的，因此非常成功。

The Cuckoo Club

Designer
Tim Mutton, Jo Sampson

Location
Soho, London

Area
465 m^2

Photographer
Suzanne Mitchell, Edmound Sumner

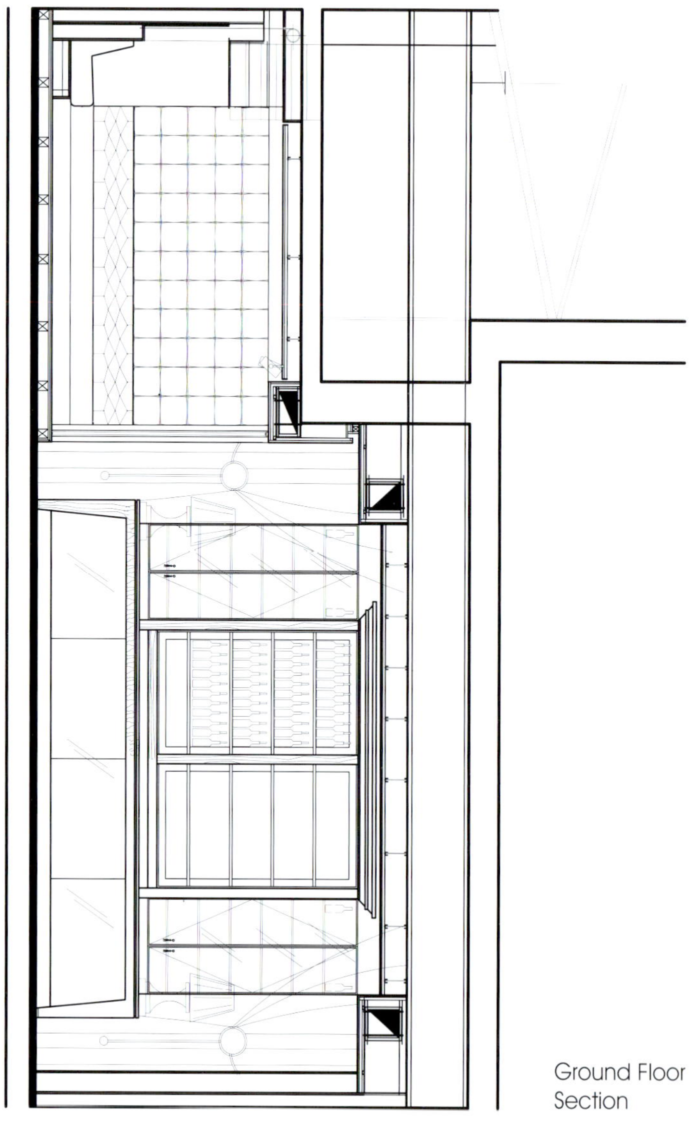

Ground Floor Section

The Cuckoo Club is a west-end venue with a difference. With a brief to appeal to London's party animals tired of fast-burn, cavernous and faceless venues, The Cuckoo Club has become the hottest club in London, successfully marrying real old-school glamour with a completely modern context for a wide-ranging clientele, from Bryan Ferry, Robert de Niro and Prince Harry to Kate Moss, Paris Hilton and fashion designer Julian McDonald.

Both a nightclub and destination restaurant, The Cuckoo Club is located on the former site of the Stork Rooms in Swallow St, London W1: once a revered staple of London's nightlife, favoured by bohemians and aristocracy alike. The space was stripped out and completely reconfigured to meet the clients' vision, taking out two mezzanine floors, removing the existing stairs and creating a brand new timber-clad steel structure double-height staircase plus reinforced concrete floor areas.

Blacksheep envisaged The Cuckoo Club as a kind of filmset, creating a grand, dramatic, "cigar box" fantasy house with timber cladding, huge doors, sweeping stairs, silk and voile drapes, endlessly reconfigurable LED lighting and lots of fun, original, statement design features, from a gold padded timber-framed alcove on the ground floor to a bar entirely covered in gold sequins on the lower ground.

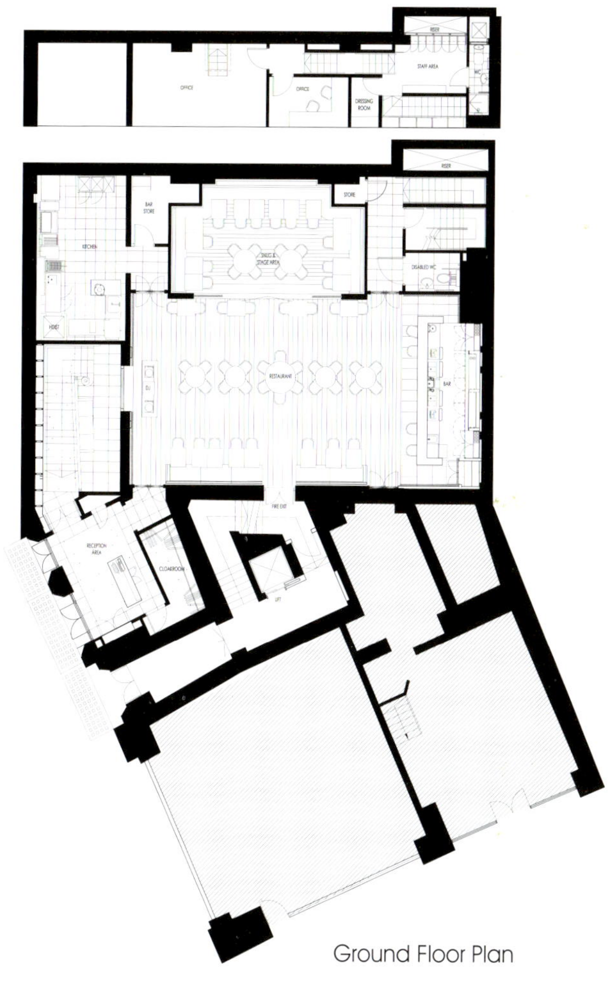

Ground Floor Plan

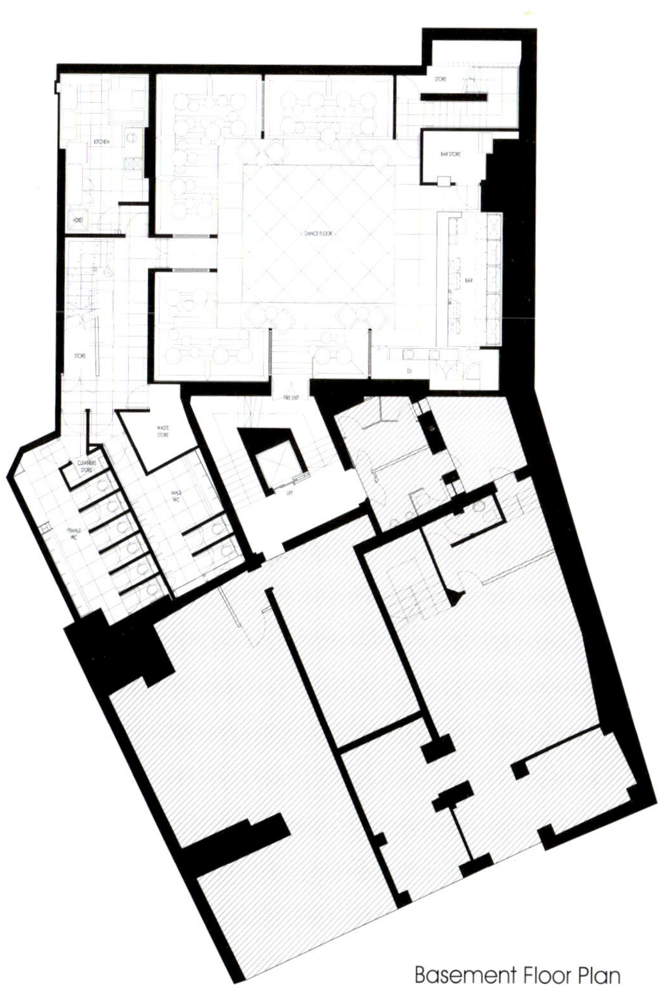

Basement Floor Plan

Cuckoo 夜总会是一个与众不同的西区场所。该设计的初衷是为了吸引伦敦众多厌倦了快热、空洞、无个性场所的派对玩家。Cuckoo 夜总会已经成为伦敦最热门的聚会场所，它成功地将真正的老派魅力与完全现代的环境结合在一起，吸引了从布赖恩·费瑞、罗伯特·德尼罗、哈里王子到凯特·莫斯、帕丽斯·希尔顿和时装设计师朱利安·麦克唐纳等不同的客户群。

Cuckoo 夜总会既是夜总会又是招牌餐厅，位于伦敦西区 Swallow 大街 Stork Rooms 的旧址，这里曾是备受推崇的伦敦夜生活的标志，颇受波西米亚人和贵族青睐。其原先的空间被完全拆除，重新装修，以满足客户的愿景，拆除了两个夹层以及原有的楼梯，创建了一个全新的、外部由木材包覆的钢结构建筑，建造了两层高的楼梯，地板由钢筋混凝土铺设。

Blacksheep 希望将 Cuckoo 夜总会设计成一个拍摄布景，打造一间豪华、梦幻、类似“雪茄盒”的房子，拥有木质外壳、巨大的门、干净的阶梯、用丝绸或薄纱制成的窗帘、可不断变换重组的 LED 灯光，传达出无尽的欢乐、创意，从一楼由金色填充物填充的木框壁龛到底层镶嵌着金色亮片的吧台，到处都是设计的亮点。

Francois Frossard Design

SET Nightclub

Designer
Francois Frossard

Location
Miami, USA

Area
929 m^2

Francois Frossard crammed a whole lotta style into the venue's 929 square meters, creating what he calls "a 1940s Hollywood mansion ambiance". The dance-floor-less space boasts working fireplaces, decorative elephant tusks, authentic Pucci fabrics, and two tube-shaped, floor-to-ceiling, glass-enclosed pneumatic elevators for dancers. Even the usual disco ball gets replaced by four spherical chandeliers in glittering Swarovski crystal. The resulting look is richly layered and full of glamour. Frossard custom-designed each piece of furniture, and also had a hand in creating the lighting and video systems.

SET features a "euro-hip" go-go style interior. Francois made sense of a property known for notorious epic failures. Francois understood the mistakes of his predecessors and drew on Miami's "modern" European style. Although SET is majestic it is free of pretension and a satire of the buildings' former lives. Francois installed a working glass fireplace; each piece of furniture is covered in luxurious Pucci fabric and the Chihuly inspired chandeliers mock authority.

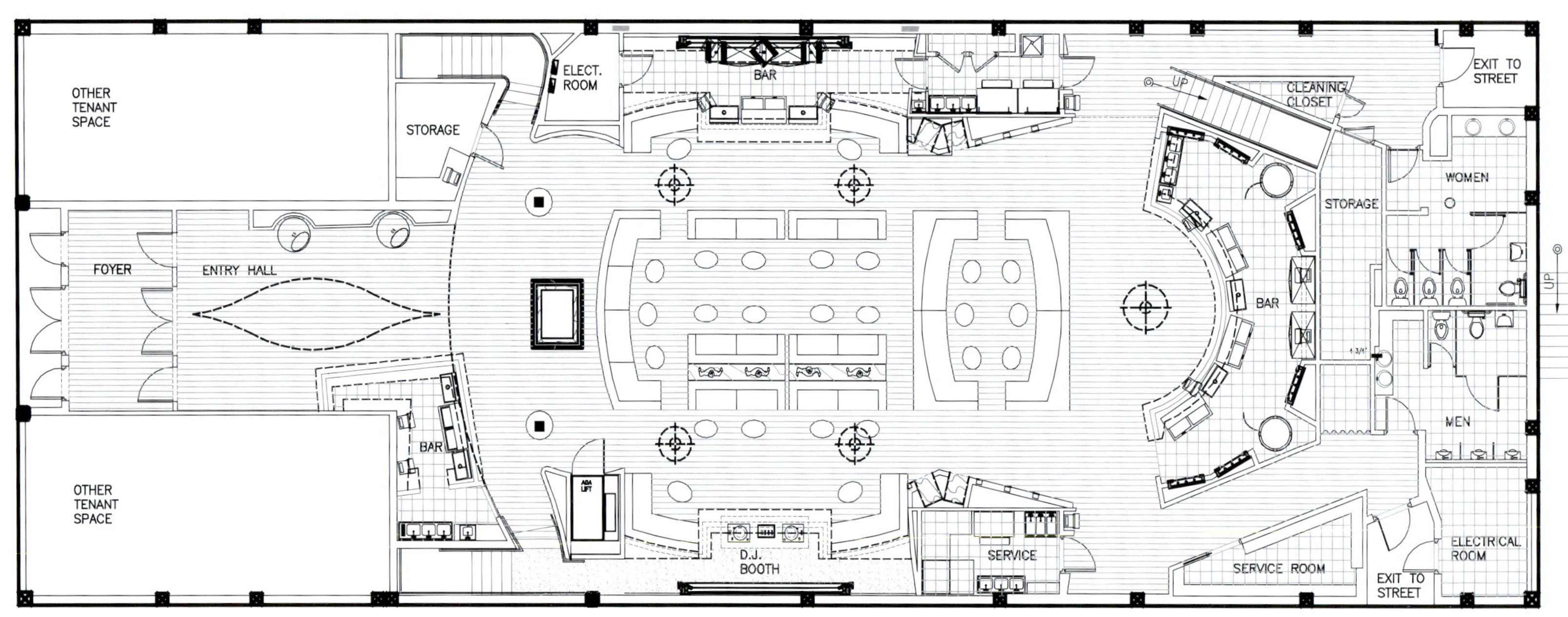

FIRST LEVEL FLOOR PLAN

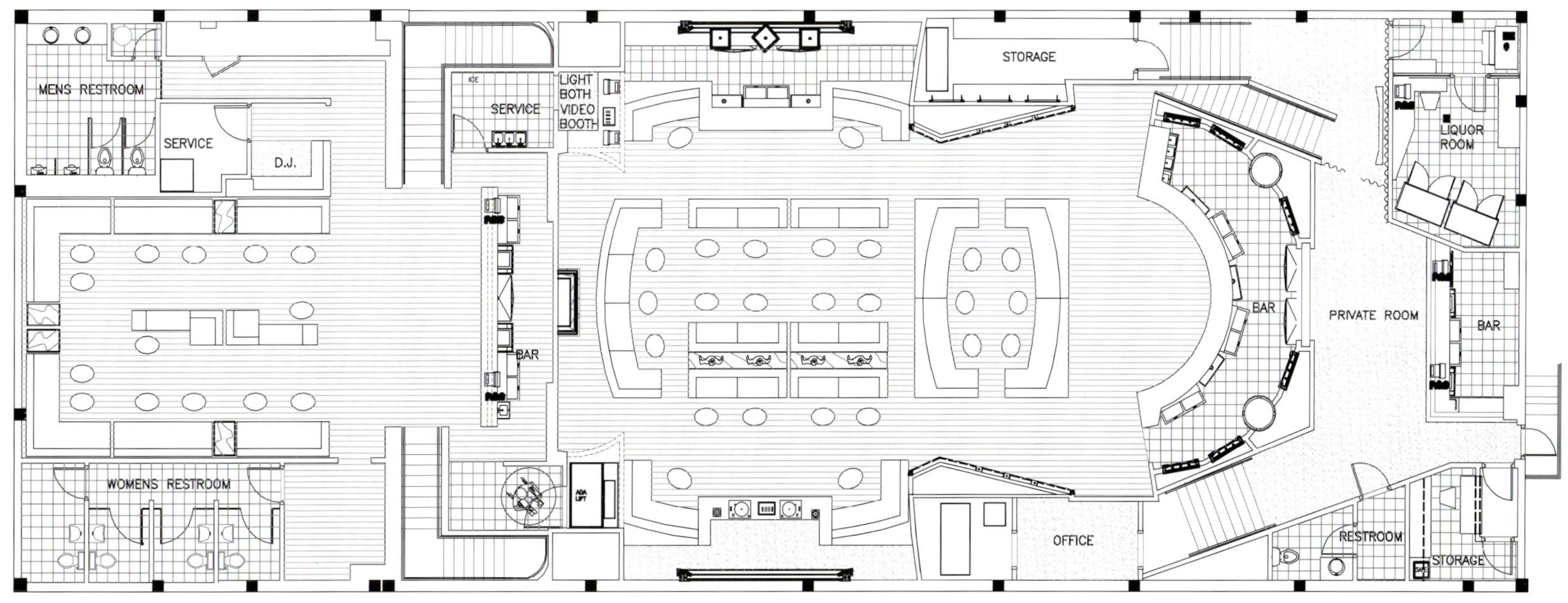

SECOND LEVEL FLOOR PLAN

该夜总会面积为 929 平方米，在室内装修中，Francois Frossard 运用了多种装修风格，营造出了他想要的“一种 20 世纪 40 年代的好莱坞豪宅氛围”。夜总会没有舞厅，却因可使用的壁炉、装饰象牙、正宗的普齐织物和两个用于跳舞的管状落地玻璃电梯而闻名，甚至连惯用的迪斯科闪光灯球都被替换掉了，取而代之的是四个镶嵌着闪闪发光的施华洛世奇水晶的圆形吊灯。最终，设计看起来层次丰富且充满魅力。夜总会的每一件家具都是 Francois 设计定做的，同时他还参与了照明与视频系统的设计。

SET 的特色在于其“euro-hip”摇摆舞风格的室内设计。Francois 清楚过分铺张、讲究排场会导致失败，也明白了先前设计师失败的原因，因此，他采用了迈阿密现代欧式风格来装饰空间。尽管 SET 很宏伟，但是却不张扬，也充分显示了对先前居住者的尊重。Francois 为其安装了一个实用性的壁炉，每件家具外都罩上了奢华的普齐织物，其上还挂有仿奇胡利风格的吊灯。

Fred Mafra

Roxy Josefine

Designer
Fred Mafra

Area
955 m^2

Photographer
Jomar Gragança

The sculptural forms are now based on straighter lines and abandon the traditional 90-degree angle.

The key shape adopted was the hexagon and the exploitation of all its faces, in different sizes. Prismatic elements, pillars formed from the union of triangles, squares and other lapidary figures instigate and transform one's perception depending on where it stands along the dance floor. The asymmetry of the design appears in every detail.

The facade is now dark black. The glass walls visible from inside were coated with dark-silver glassy material. The internal structure that holds the facade together from inside is built up in triangle shapes made out of stainless steel. Each glass module above the entrance aisle received a LCD panel. In total there are 20 LCD panels facing the outside of the club.

The DJ booth which was placed at the center of a stage is positioned at the end of the dance floor, with a full view of the club. The dance floor is a long aisle that stretches itself for 30 meters long and 6 meters wide. It is now three times larger than the previous one. In one of its sides is the main bar, extended along its length. It is composed of seven bays that divide custom care and avoid crowds, which are a common problem for linear bars. The bays were framed into trapezoidal porticoes, built with double drywall and acoustic system that reduces the entry of sound, allowing the bartender listens perfectly customer's request at each niche. These gateways also hide the modular structure of the pillars supporting the estruture.

This four tonnes structure is composed of over 100 hexagons. Each of these hollow hexagons serve as a protective for the LED light connected to a video pixel mapping panel control. The aluminum structure in "U" receives tapes LED's inlays and then sealed with polycarbonate.

The result is a textured ceiling, three-dimensional lights in constant motion, teasing you to leave reality behind. It's like stepping into a video clip with real graphics and lights that dance to the music.

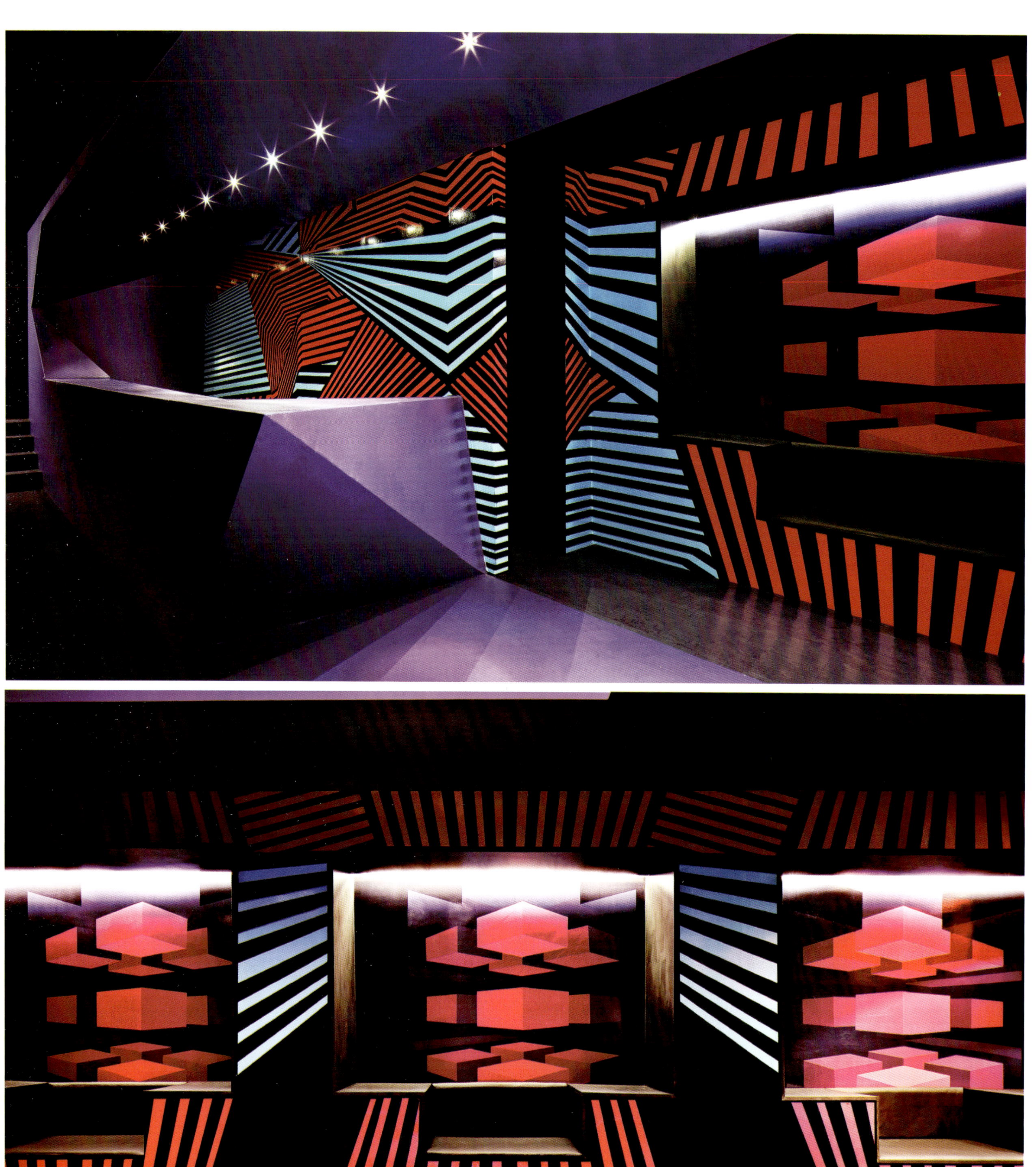

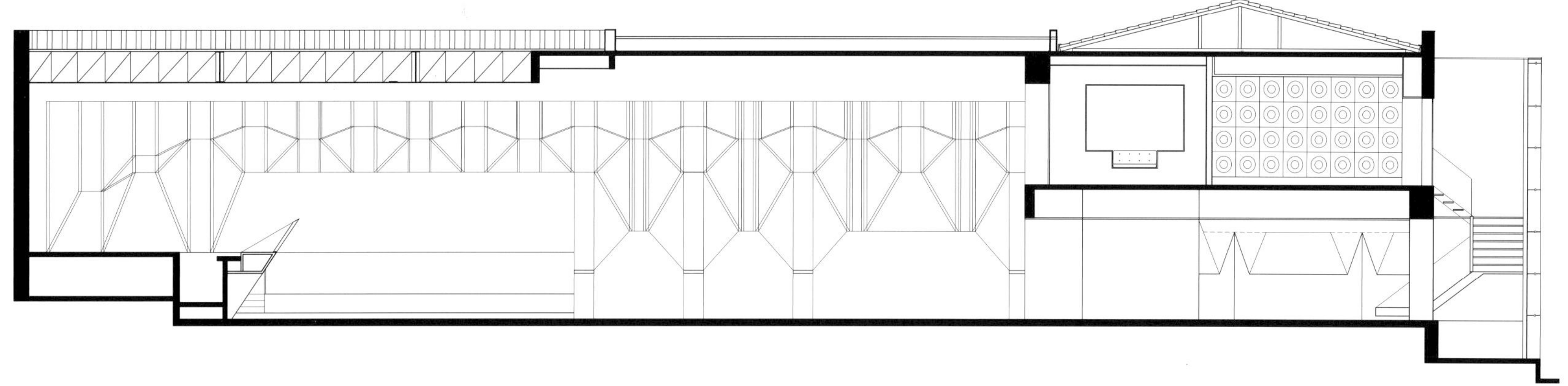

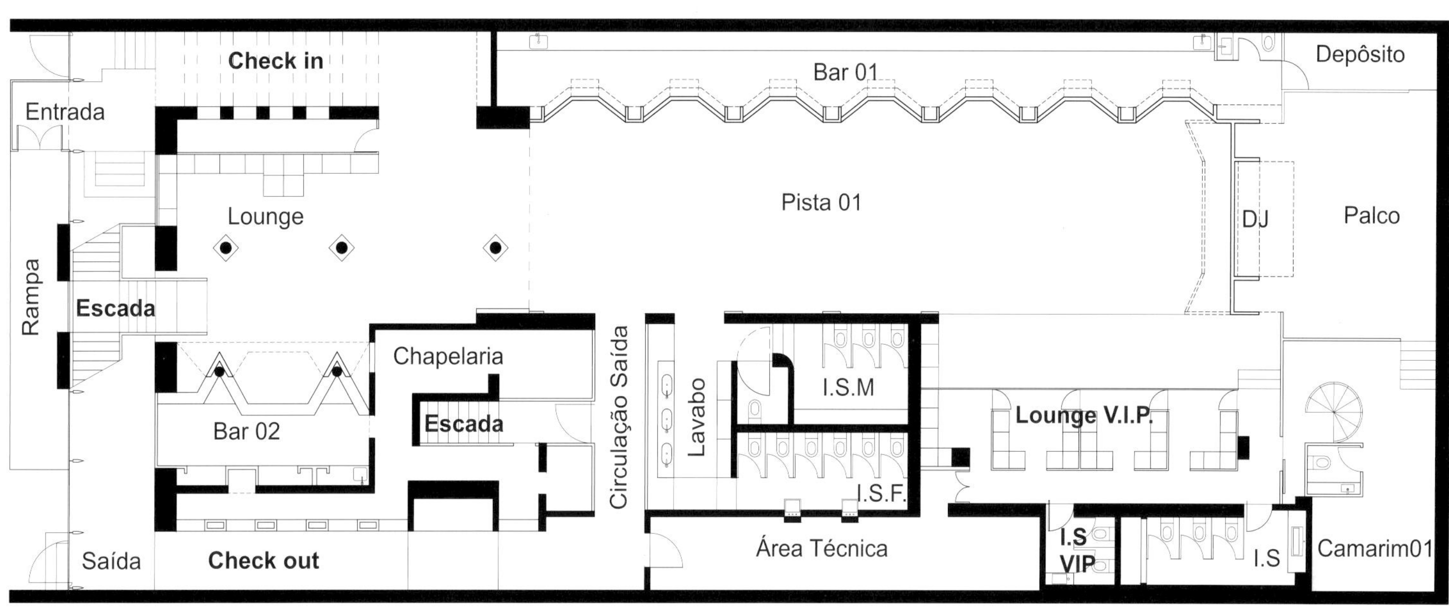

Check in
Entrada
Bar 01
Depôsito
Lounge
Pista 01
DJ
Palco
Rampa
Escada
Chapelaria
Circulação Saída
Lavabo
I.S.M
Lounge V.I.P.
Bar 02
Escada
I.S.F.
Área Técnica
I.S VIP
I.S
Camarim01
Saída
Check out

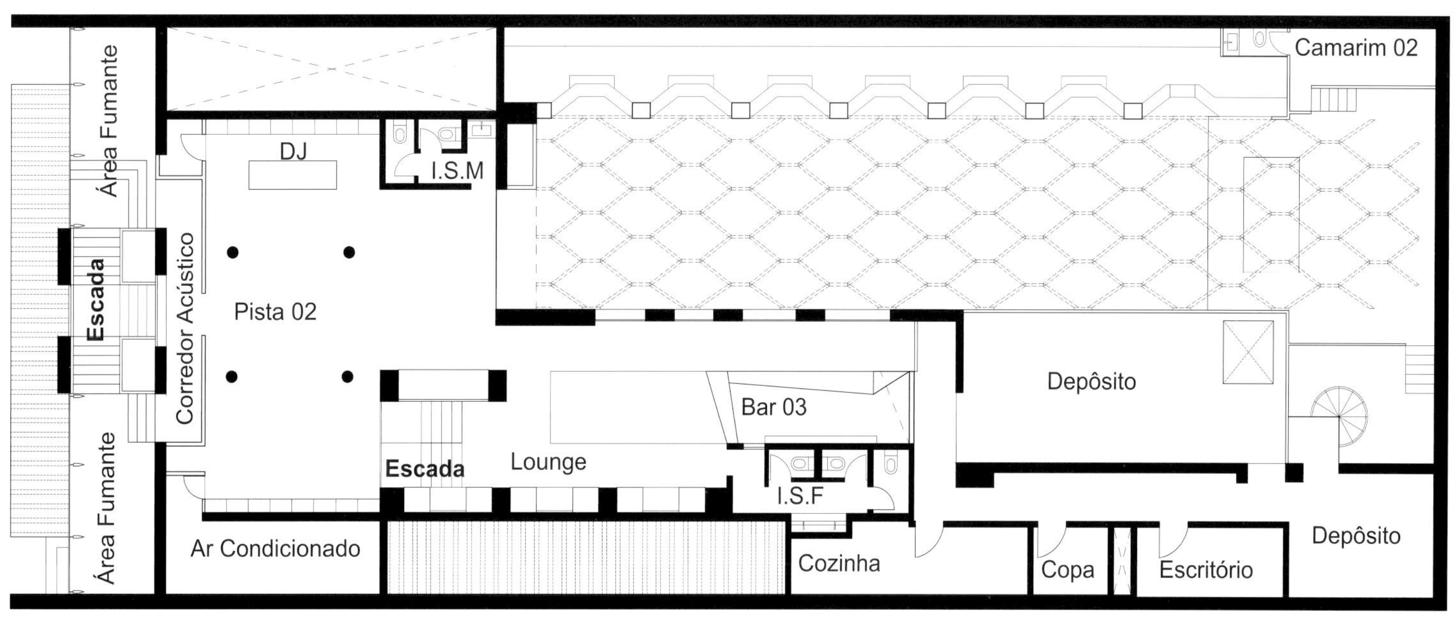

Camarim 02
Área Fumante
DJ
I.S.M
Escada
Corredor Acústico
Pista 02
Depósito
Bar 03
Escada
Lounge
I.S.F
Área Fumante
Ar Condicionado
Cozinha
Copa
Escritório
Depósito

建筑的造型采用了更直的线条，放弃了传统的 90 度角。

设计主要运用六边形，利用了不同大小的六边形及其六个面。棱柱元素，由三角形组合而成的柱子、正方形和其他宝石的形状，这些元素散布于舞池各处，煽动着人们的情绪，也改变着人们对空间的认知，这种不对称的设计出现在每一个细节中。

现在的外立面是深黑色的。从室内看去，玻璃墙壁仿佛被涂上了一层暗银色的闪光材质。三角形的不锈钢构成了整体内部空间以及外立面的支撑结构。入口上方的每个玻璃模块上都装有液晶显示屏。夜总会的外面总共装有 20 块液晶显示屏。

DJ 台设置于舞台中心，位于舞池的尽头，能够一览夜总会的全景。舞池是一个长长的通道，长 30 米、宽 6 米，现在的舞池是以前的三倍大。舞池的一侧是主吧台，沿着舞池的长边而设。设计还设置了七个隔间，便于服务顾客，并可避开人群，这种设计在线性酒吧中很常见。隔间整体形状类似阶梯门廊，由双层的石膏板和隔音设施构成，能够尽量减少外界的声音，让酒吧服务员能够听清楚顾客的要求。建立双墙和隔声系统，减少了声音的输入，让酒吧服务生能在每一个隔区里完全听到客户的要求。这些门廊式设计还有助于隐藏支撑着整体结构的模块化的柱子。

这个重达四吨的柱廊结构由 100 多个六边形构成。作为 LED 灯的保护装置，每一个空心的六边形都连接着一个视频像素图控制面板。铝结构的 U 形接收面与 LED 的内嵌物相连接，外面则由聚碳酸酯密封住。

最终，建筑充满纹理质感的天花和不停变换的三维灯光，会让顾客忘却现实，仿佛走进了一个拥有真实影像和灯光效果的视频片段中，随着音乐翩翩起舞。

Lervik Design

PUSH Nightclub

Designer
Alexander Lervik

Location
Gothenburg, Sweden

Area
600 m^2

PUSH Nightclub is the latest design by the founders of such illustrious clubs and restaurants as Hell's Kitchen, The White Room and Spy Bar. True to form, both the interior and exterior of PUSH reflect Lervik Design's dedication to dramatic, bold architecture that's solidly grounded in a modern sensibility. The focus is a stunning series of contrasting elements as you move through the space, which has become a staple of Gothenburg nightlife.

The entrance to PUSH features deep purple and red lighting against strong lines and reflective surfaces, topped off by a surprisingly delicate web of gossamer lighting. Moving through PUSH, you'll appreciate the successs of Lervik Design's modern nightclub aesthetic, which successfully sets an intimate mood in a large space by breaking it down into smaller areas that encourage intimacy in a large-scale setting.

The nightclub design offers bold white sofas and black tables, soft lights against a stark white stretch of bar festooned with sculpted swags, and the unexpected beauty of an single, enormous mosaic rose on the floor all combine flawlessly to create a dreamlike, moody atmosphere.

PUSH 夜总会是一项最新设计，它的设计师曾设计过很多著名的夜总会和餐厅，例如，Hell 厨房、白色空间和 Spy 酒吧。一如既往，PUSH 夜总会室内和室外设计都反映出了 Lervik 设计艺术大胆的风格，最终，夜总会沉稳地立于现代都市之中。游走于空间之中，你会发现一系列具有鲜明发差的元素，而它们则是 PUSH 夜总会的标志。

PUSH 夜总会的入口以深紫和红色的灯光与强烈的线条感和反射面的对比为特征，形成了一种惊人的微妙的薄纱照明网。游走于 PUSH 夜总会，你会体会到 Lervik 设计的现代夜总会的成功之处，这种设计通过将大空间分解成多个小区域，成功地在一个大空间内营造出亲密的氛围，有助于在大空间背景下营造亲切感。

夜总会设计选用了醒目的白色沙发和黑色桌子，柔和的灯光下，光洁的白色吧台与雕刻的悬挂物交相辉映。地板上，一个巨大的马赛克玫瑰图格外美丽。所有的元素完美地组合在一起，共同营造了一个梦幻仙境。

Provocateur Lounge & Nightclub

Designer
Loinel Ohayon, Siobhan Barry

In the heart of the Meatpacking District, beyond an iron fence, and through a darkened entry is Provocateur – a space where 18th century gardens and drawing rooms meet 21st century sound studios and ultra lounges, creating an intoxicating stage set for New York nightlife.

The space is divided into 2 areas – an outdoor garden lounge and a high-energy nightclub.

Dramatic dollhouse inspired entry drapery opens up to a whimsical picturesque center garden with tall birch trees, hanging iron bird cages, candle holders, stone carved statues, hanging swings and benches filled with romantic multi-colored floral fabric.

Private staged "rooms" surround each table creating unique surreal scenes that deviate from reality exuding an oddly playful, naughty quality where the guests bring the sets to life as if they are a part of a photo shoot.

Through the garden lounge is a high energy house-music nightclub. Custom-made metal wings hover over the main bar providing the perfect backdrop for catwalk dancers. The space is filled with elegant lavender hues with a touch of gold.

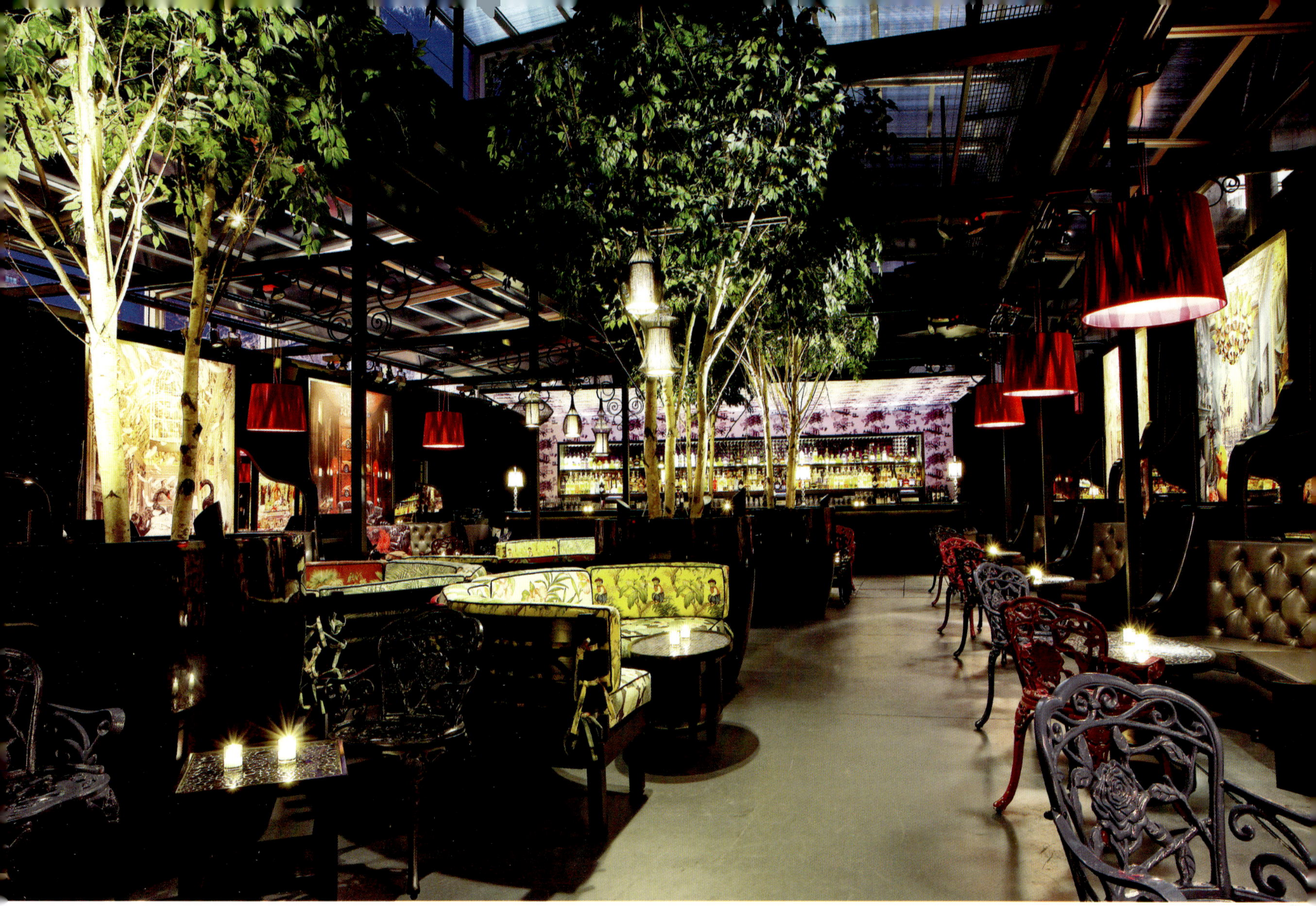

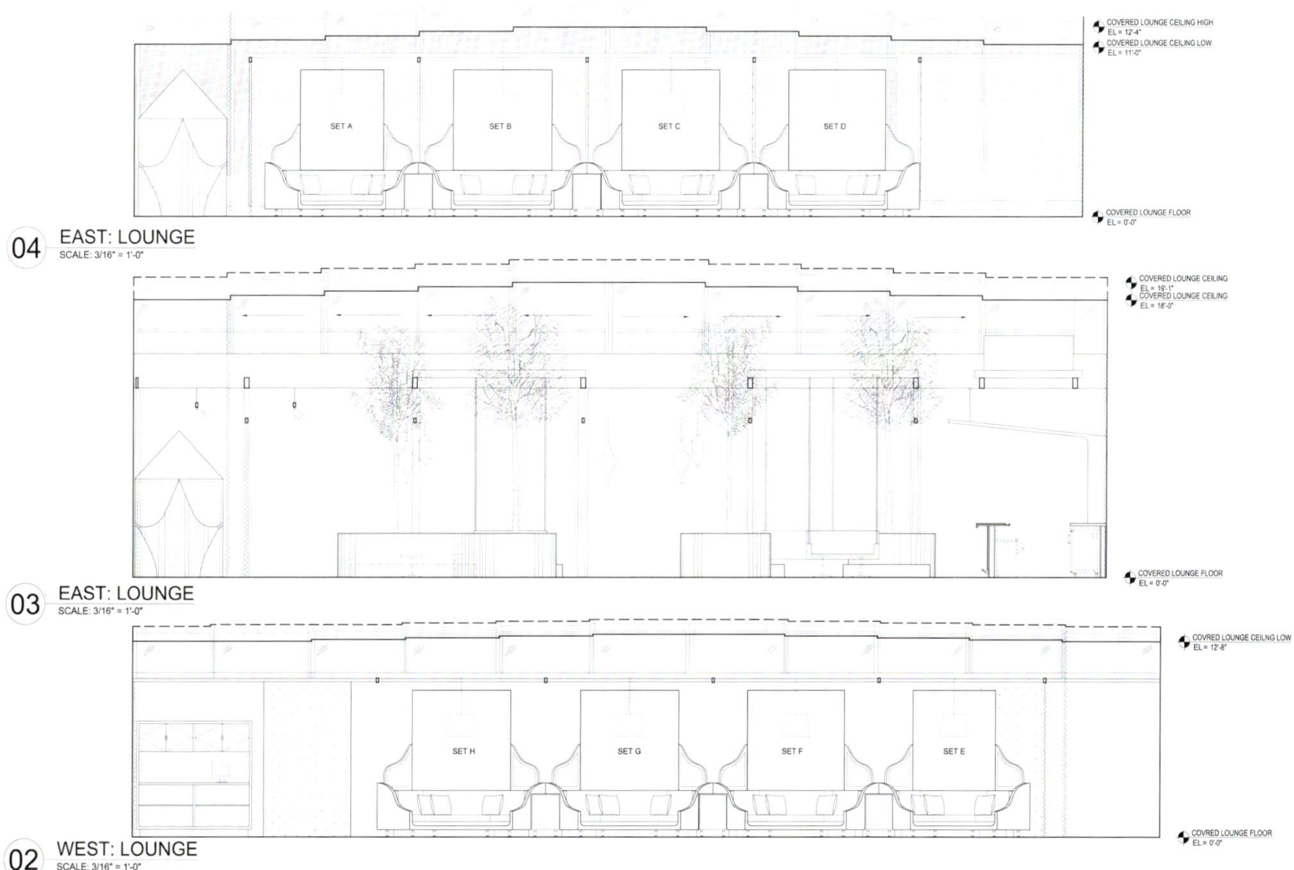
COVERED LOUNGE CEILING HIGH
EL = 12'-4"
COVERED LOUNGE CEILING LOW
EL = 11'-0"
SET A
SET B
SET C
SET D
COVERED LOUNGE FLOOR
EL = 0'-0"
04 EAST: LOUNGE
SCALE: 3/16" = 1'-0"
COVERED LOUNGE FLOOR
EL = 0'-0"
03 EAST: LOUNGE
SCALE: 3/16" = 1'-0"
COVRED LOUNGE CEILING LOW
EL = 12'-8"
SET H
SET G
SET F
SET E
COVRED LOUNGE FLOOR
EL = 0'-0"
02 WEST: LOUNGE
SCALE: 3/16" = 1'-0"

在米特潘肯区的中心，铁栅栏的另一侧，穿过一个幽暗的入口，就来到了Provocateur，18世纪的花园和客厅与21世纪的录音室和高级休闲吧相结合的设计风格，为纽约的夜生活营造出一个令人心醉的场景。

整个空间分为两个部分：室外花园休闲吧和一个活力十足的夜总会。

拉开夸张的玩具屋风格的入门帏帐，就会进入一个奇异而别具一格的中心花园，这里到处都是高高的桦树、高挂着的铁鸟笼、烛台、石头雕像、秋千、长条凳和浪漫的五颜六色的花布。

每张桌子周围都围绕着一个个私密“房间”，营造出一副与现实格格不入的超现实主义的场景，散发出一种颇为怪异、好玩而又顽皮的气质，而客人们似乎是照相机前的明星，给整个场景注入了生机。

穿过花园休闲吧，就来到了富有活力的浩室音乐会所。特别定制的金属翅膀悬挂在主吧台的上方，给猫步舞者提供了一个完美的背景。整个空间充满了优雅的淡紫色调，间或夹带一抹金色。

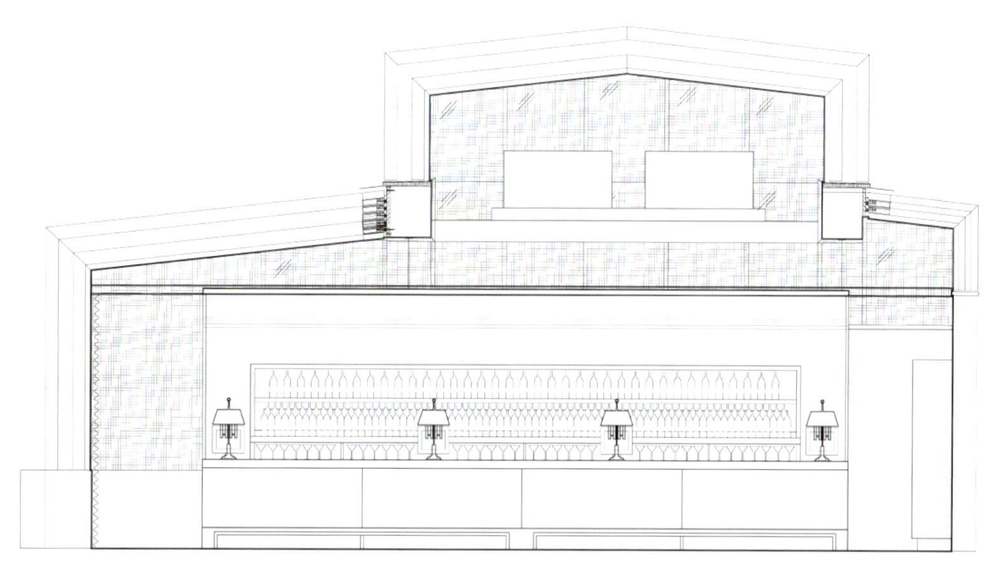

05 SOUTH: LOUNGE BAR
SCALE: 3/16" = 1'-0"

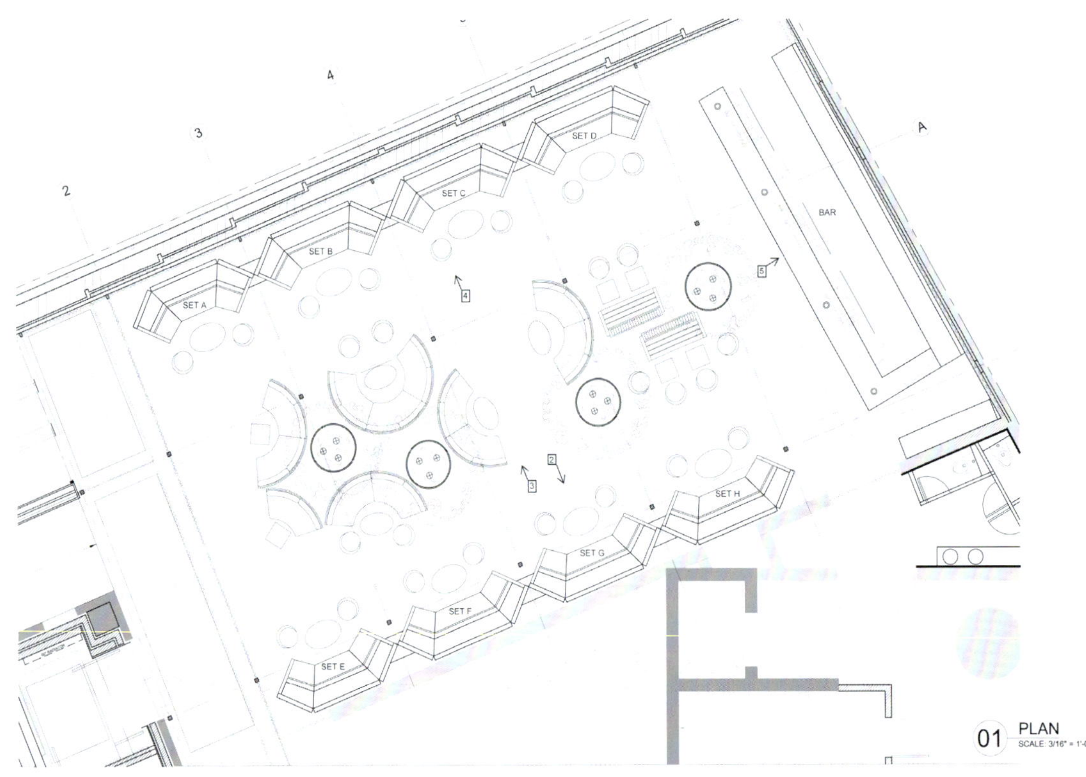
2
3
4
A
SET A
SET B
SET C
SET D
SET E
SET F
SET G
SET H
BAR
01 PLAN
SCALE: 3/16" = 1'-0"

BAR
BAR

酒吧 BAR

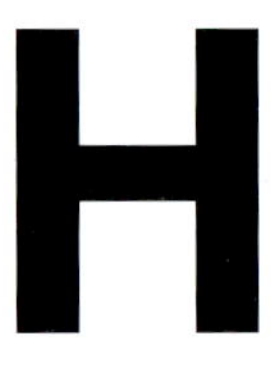

orizon Rooftop Restaurant & Bar, Hilton Pattaya

Designer	Location	Area
dwp | design worldwide partnership	Pattaya, Thailand	800 m^2

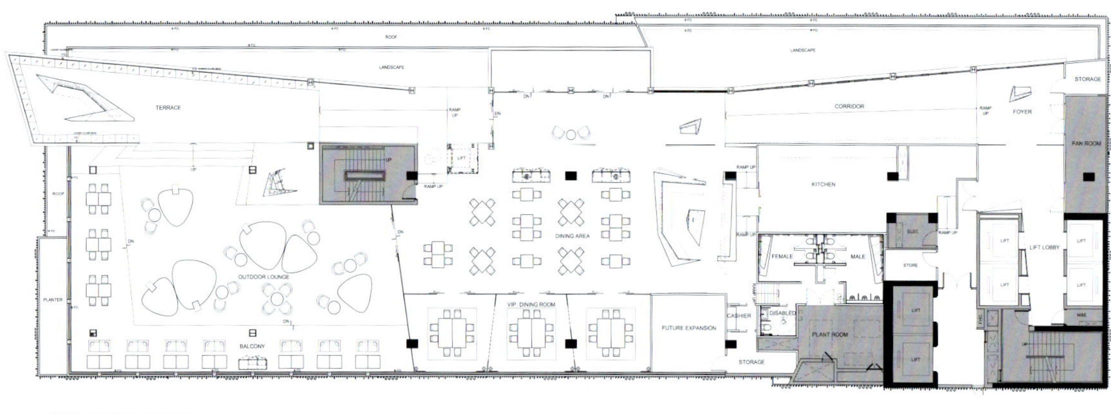

Set over 800 m^2, a truly unique and innovative rooftop concept, the new restaurant and bar sat high atop the Hilton Pattaya is a world unto its own. Offering dramatic views of the city by the sea, guests feel transported to their own exclusive clifftop pedestal high above the world. dwp created an idyllic place for chilled out tunes and the latest professionally concocted drinks and eats.

A swift lift ride from ground level to up on high, guests can amble slowly up the grand staircase, overlooking the enigmatic views of the ocean. Whether dining on the lower level, or sipping champagne above, this unparalleled experience in Pattaya is set to breathtaking heights, awe-inspiring décor and eye-catching light displays. Transforming this rooftop, which was not originally intended for F&B use, was challenging, given the limitations of the space, existing M&E equipment housed there, and requirements for such a venue. By including meticulous space planning, injecting sleek, clean lines and capturing the extended space beyond the normal footprint of the building with the suspended "Infinity Bar", dwp exceeded client expectations.

One of the most remarkable architectural aspects, taking outdoor dining to the next level, is the open-air "Stargazer Lounge", where guests can literally witness the relative movement of the stars through the circular opening in the overhanging roof. Indoor and outdoor dining spaces blend seamlessly together, with large expansive windows indoors to capitalize on the views. The resulting effect is one of both spaciousness and intimacy.

这家新餐厅兼酒吧坐落于芭堤雅希尔顿酒店的顶部，总面积超过 800 平方米，以真正独特和创新的屋顶概念为设计理念，形成了一个独有的世界。在这里，能够欣赏引人注目的海边城市景观，客人会感觉像穿越到专属于自己的悬崖顶部，屹立于世界之巅。dwp 创建了一个世外桃源，让人们享受放松的音乐，品尝由专业人士最新调制的饮料和食物。

乘坐电梯可以快速从底层到达顶楼，当然，客人也可以悠闲地漫步于豪华的楼梯上，一边走、一边俯瞰神秘的海景。在芭堤雅，无论是在低楼层中进餐，还是在高层上饮酒，这无与伦比的体验都被设定在激动人心的高度、令人惊叹的装饰和引人注目的照明设备上。屋顶本来没打算作为餐饮部使用，因为此处空间限制较多，原本的机电设备也安置在此处，因此，在这样一个场所建造餐饮部，是非常具有挑战性的。通过细致的空间规划、流畅、简洁线条的应用以及对建筑扩展空间的把控，dwp 建造了这个悬浮的“无限酒吧”，超额完成了客户的期望。

本案中，户外“观星酒吧”的设计将室外餐饮推向另一高度，是最卓著的建筑策略之一，透过屋顶上的圆形开口，客人可以观看星体的相对运动。室内和室外就餐空间完美地交织在一起，室内开阔的大玻璃窗使客人可以观赏室外景色，进而形成既宽敞又亲密的空间。

New Kuvee Discothèque

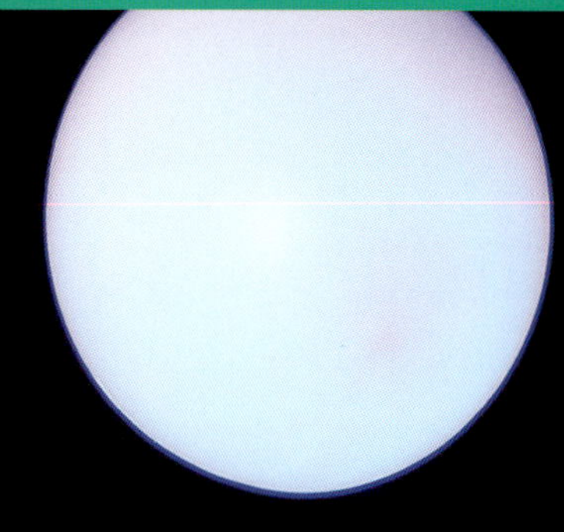

Graphic Design
Elia Felices Interiorismo

Location
Barcelona, Spain

Area
920 m²

Photographer
Rafael Vargas

New Kuvee Discothèque is projected around the space and futurist concept. On this way, the visitor, through his experience, is transported to a far place and time.

Enter to the discotheque through the industrial unit's courtyard. Immediately you can find the hall, where you can observe a diaphanous place with light balls suspended and a volatile window which works as the wardrobe place. Three circular bars composed by futurist and retro illuminated shelves are distributed. The shelves are placed in different heights, which make the sensation if they were playing to interlace. The main bar, which is situated in the middle, is joined with the disc-jockey's cabin creating an orbit with the two left bars. The other ones are located near the two chill-out areas where floated graphics and projected images take place.

Through the continuous white floor, you arrive to the wardrobe place, which call you through its focal light and gives you welcome. The graphic curves of the wardrobe slide through the walls and wait us to leave again. A large zone shows you different areas, each one with a special character: firstly you are introduced to a large dancing floor and find yourselves around of moving bars, podiums, lights and graphics. This dancing floor is the center where you let yourselves go to the quietest chill-out zone. In front of them there are the circular bars which are made of metal sheet. In addition, we can see the entwined bottle racks like musical notes in a pentagram.

Finally, a white neutral curtain gives way again to the light and color dynamism, and the space is transformed one and another time giving way to new perceptions, which are felt with the five senses.

SECCIÓN C-C'. NEW KUVEE DISCOTHÈQUE

SECCIÓN D-D'. NEW KUVEE DISCOTHÈQUE

New Kuvee 迪斯科舞厅围绕着空间和未来主义的概念而设计。通过这种方式，光顾这里的顾客，会有被传送到某个遥远的空间和时间中的感觉。

穿过工业园区的院子，你马上就会发现迪斯科舞厅的门厅，门厅设计精致，头上是悬吊着的光球，窗户被改造成了衣柜。三个由未来主义和复古风格的架子组成的吧台，分散在各处。架子有高有低，给人一种即将相交的感觉。主吧台位于中间，配有 DJ 台，与其余两个吧台连成一线。剩下的两个吧台靠近两个休闲区，这里有悬浮图形和投影图像。

沿着连续的白色地板，你会到达衣柜处，此处，聚焦光呼唤着你并给予你热烈的欢迎。曲线图形的衣柜可随墙壁滑动，静候顾客再次光临。一个宽敞的区域向你展示着不同的分区，每个分区都有各自独特的特征：首先，你会来到一个大型的舞厅，周围有移动吧台、乐队指挥台、各色的灯光和投影图案。这个舞厅是中心，通过舞厅便可以去往最安静的休闲区。休息区前面是由金属板制成的圆形吧台。此外，你还会看到盘旋的瓶架，像盘踞在五角星形里的音符一般。

最后，白色的中性色窗帘再一次突出了灯光和色彩的活力，空间一次又一次的转换，突出了五种感官体验的新观念。

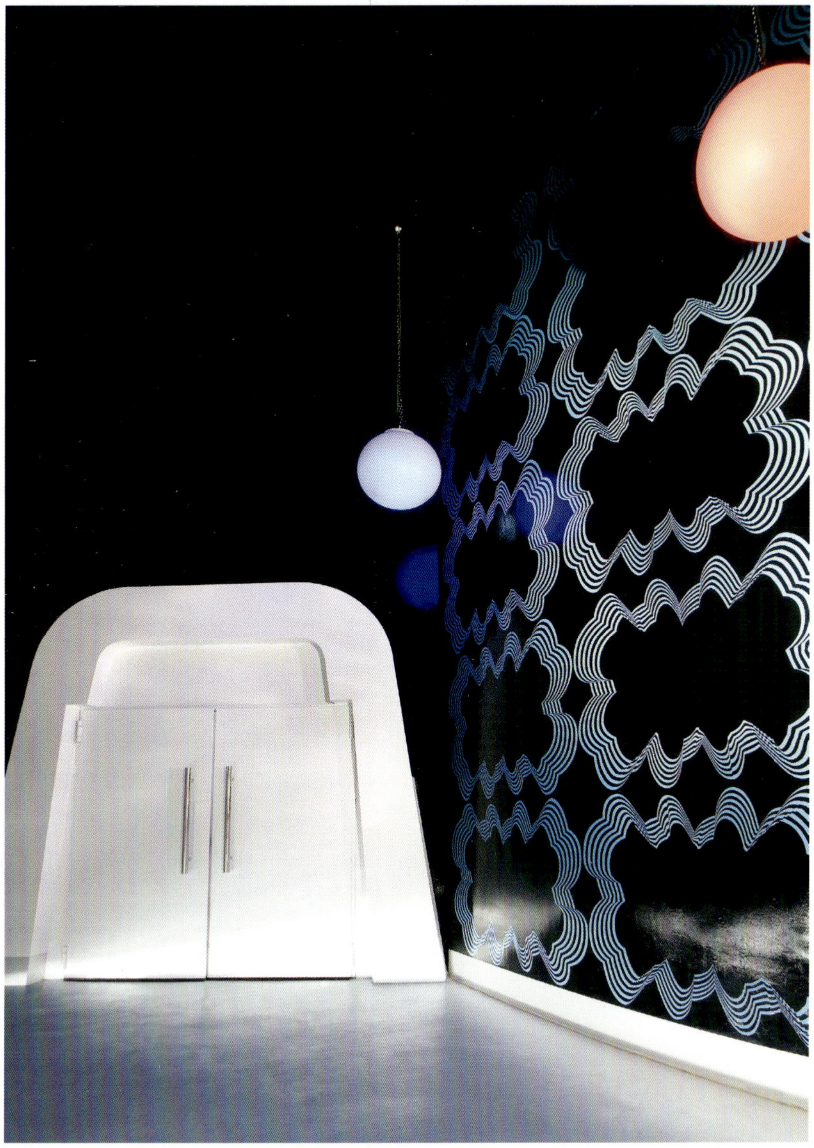

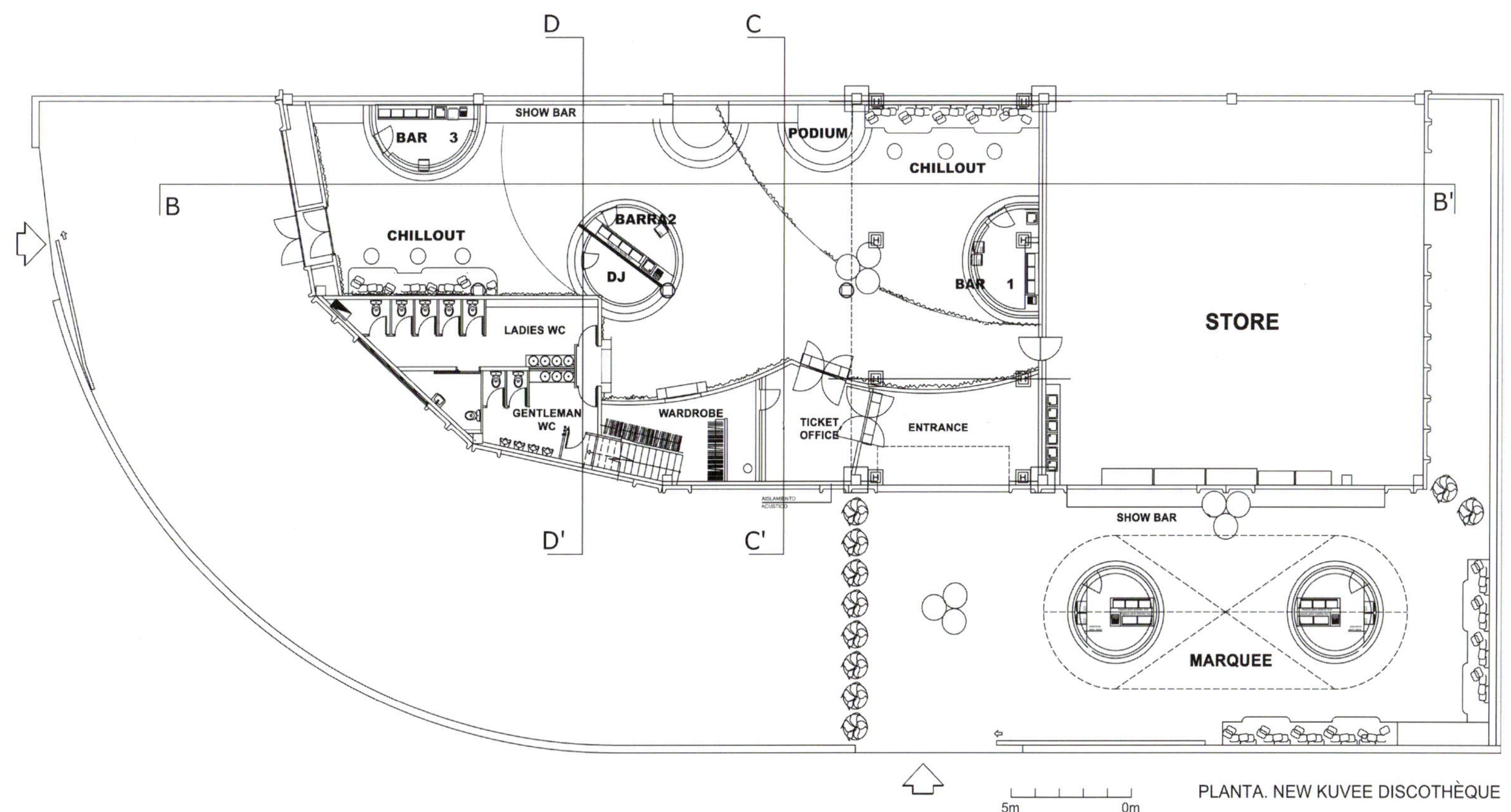

PLANTA. NEW KUVEE DISCOTHÈQUE

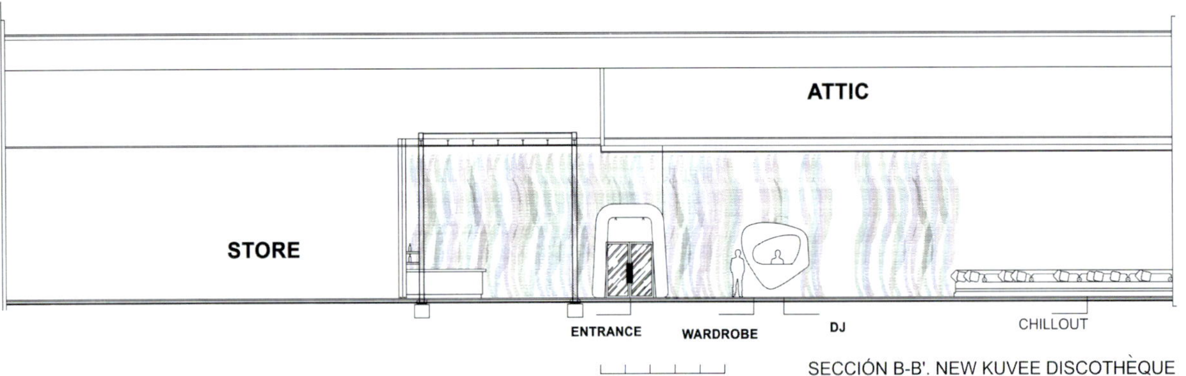

SECCIÓN B-B'. NEW KUVEE DISCOTHÈQUE

La Fuente, Music Club & Coctails Bar

Designer
Kryštof Blažek

Area
174 m^2

Location
Czech Republic

Photographer
Jan Kuděj

Kryštof Blažek created interior design for music and coctails bar La Fuente in Prague, Czech Republic. Blue-violet concrete floor in the main area is smoothly contrasted with pastel green seating and nine meters long coated bar in the same color. Black painted ceiling and walls covered by classical and a bit op-art wallpapers makes visitors feel very cozy. Dimmed light supports unique atmosphere of the space. Vertical lines, black and white curved lined wallpapers on the walls, magenta concrete floor and champagne seating make the lounge room very pleasant. Mix of traditional and modern elements creates harmonious and charming interior of La Fuente.

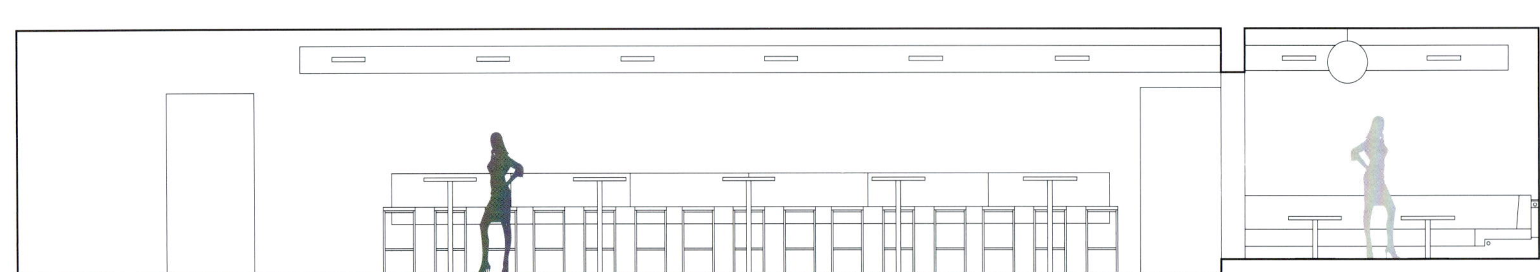

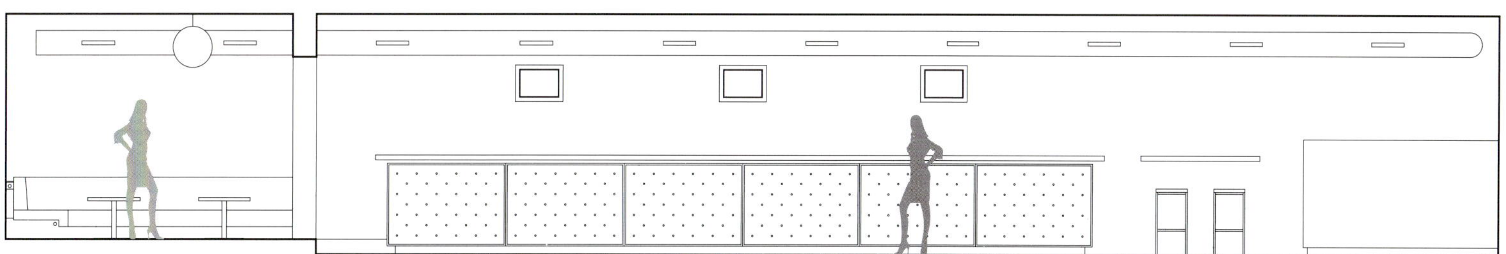

Kryštof Blažek 为位于捷克共和国布拉格的音乐鸡尾酒酒吧 La Fuente 进行了室内设计。酒吧的主要区域中，蓝紫色的混凝土地板与柔和的绿色座位和同色调的长达 9 米的吧台形成了自然的对比。漆成黑色的天花板、装饰着古典且有点欧普艺术的墙纸的墙壁，让顾客感到非常舒适。朦胧的灯光营造了酒吧内独特的气氛。墙上壁纸的垂直线条、黑白相见的弧形线条、洋红色的混凝土地板和香槟色的座位，这些设计都使休息室内的气氛非常宜人。传统与现代元素的混搭为 La Fuente 创建了和谐且迷人的室内环境。

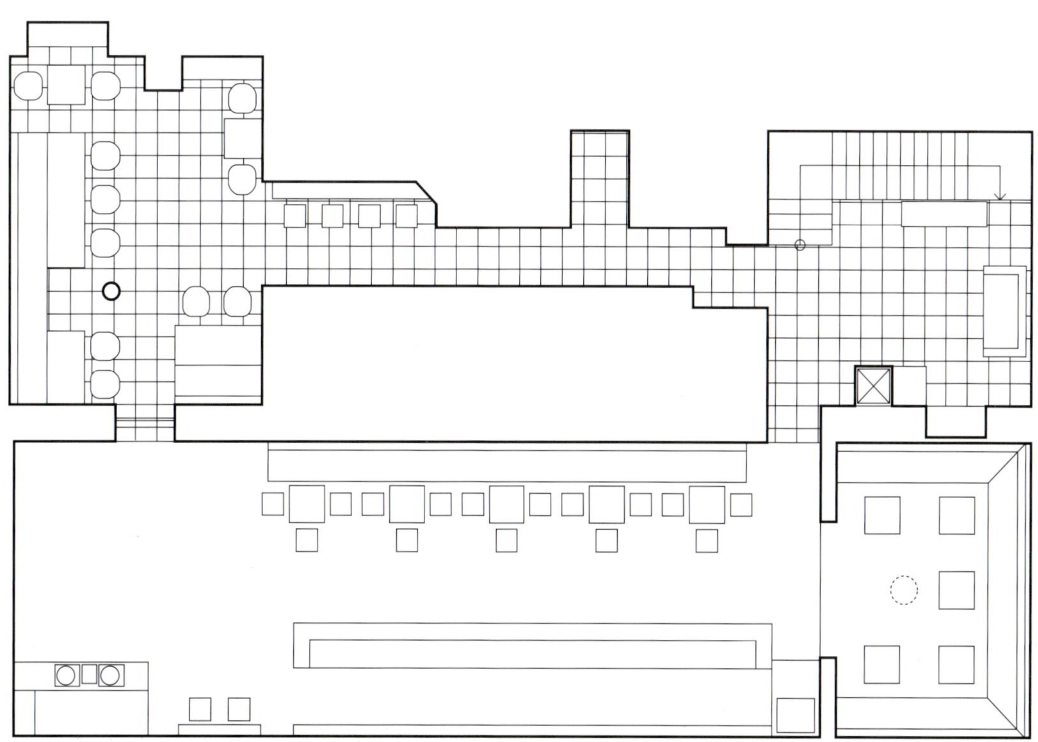

Bluarch Architecture + Interiors + Lighting

Central Lounge and Sushi Bar

Designer
Antonio Di Oronzo

Location
New York, USA

Area
372 m^2

Central Lounge and Sushi Bar is a supper club, a hospitality venue offering both food and music. The design approach was to offer the indulgent environment of a lounge and the sensual experience of the finest sushi in a rich and eclectic space.

The venue has two bars, one rectangular and one circular. Both are clad in tufted white upholstery and are topped with slabs of back-lit onyx. Above the rectangular bar, five walnut shelters curve at different heights to create a refined sense of intimacy. Each shelter has a cut-out area showing a black mirror and a Swarovski crystal sconce.

The circular section of the venue is surmounted by a mezzanine which opens into a double height topped with ivy and LED lights. The perimeter of the round area is delineated by a round banquette above which round upholstered panels in different sizes line the walls.

The experience was unique in that I was creating a venue for Japanese food without implementing the usual Japanese design language.

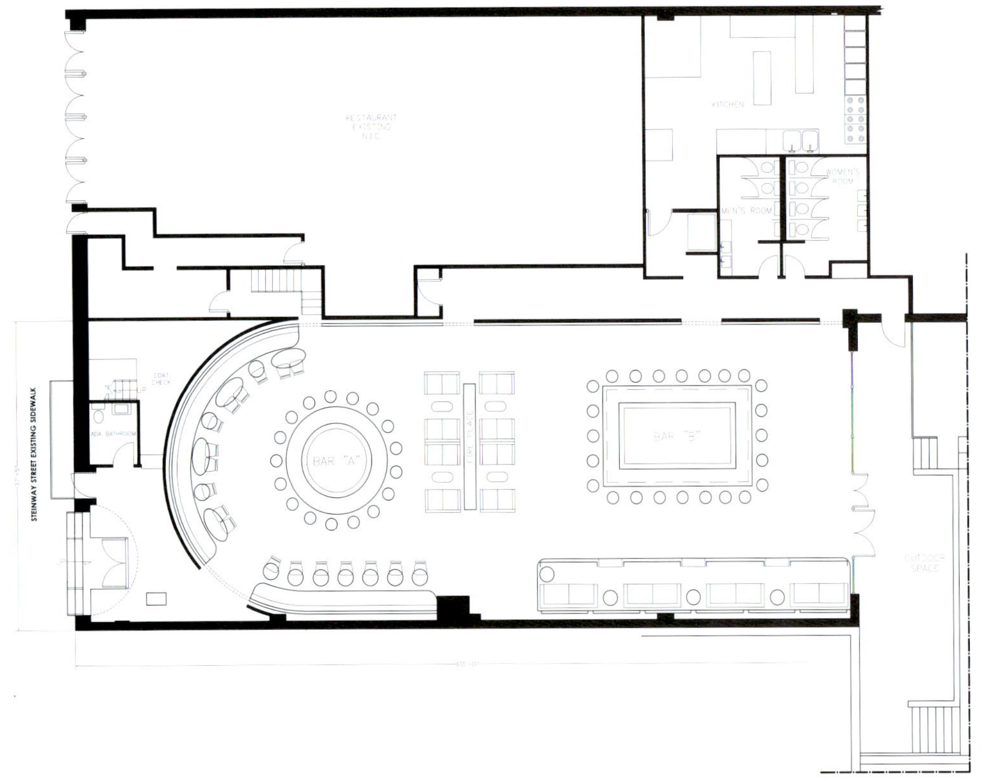

Central Lounge and Sushi Bar是一家提供食物和音乐的晚餐俱乐部及酒吧。具体设计方案是：在宽敞而颇有中性风格的空间里，营造出酒吧的放松氛围和高级寿司店的感官体验。

该项目有两种吧台，一种是矩形的，一种是圆形的。两种吧台都覆盖着簇绒白色内饰，并且上面都镶嵌着一层背光缟玛瑙。矩形吧台的上方，五条弯曲的胡桃木挡板悬浮在空中，参差不齐，恰到好处地营造出一种亲切感。每块挡板都有一个开孔区域，里面放着一面黑色的镜子和一座施华洛世奇水晶烛台。

这间酒吧的圆形区域上方是一个夹层，与圆形区域一同形成了拱顶，共两层高，顶部装饰着常春藤和LED灯。圆形区域的周边环绕着一圈圆形的座椅，座椅上方的墙上则挂满了不同大小的圆形软垫版。

这次的设计经历是独一无二的，因为在这个为日本食品所设计的场所中，设计师并没有使用常用的日本建筑设计语言。

Kokaistudios

Brownstone Bar

Designer
Andrea Destefanis, Pietro Peyron, Carmen Lee

Location
Shanghai, China

Area
130 m^2

Photographer
Charlie Xia, Kokaistudios

The "Brownstone" is a cocktail lounge located in the Surpass Court of Yonjia Lu, emerging epicenter of Shanghai night life. It was developed as a prototype for a venue of new concept in co-operation with "Blue Horizon Hospitality Group" (Blue Frog).

The project combines a simple and efficient layout, partly suggested by the limited dimensions of the venue, with the complex and redundant interplay of textures and custom-made elements concurring to define the sophisticated atmosphere and the strong identity of the place, a one-of-a-kind experience lousily inspired, in the intentions of the client, to the decadent turn-of-the-Century French "Salons".

The layout accommodates in 120 m^2 up to 80 seats of two different heights and typologies, reconfigurable in different settings and densities according to the situations and includes, a bar counter with two parallel cocktail stations, a state of the art DJ booth and a back-of-house area with a small kitchen with annex storage space.

The central part, developed around the bar counter and the DJ booth, is furnished with high seats and stools and is enclosed by a perimeter of wood finished built-in counters, same height as the bar counter, defining the edges and the access to the raised lounge platforms around it. The floor is a mosaic of small tiles of 4 different types of stone, alternatively matte or polished, creating vibrating reflections of the natural light.

BROWNSTONE

The raised lounge areas offer a lower and more comfortable seating option, with leather armchairs and sofas organized in 6 clusters around low stone-finished coffee tables.

Floor and fixed furniture are finished in dark wood, the walls in red bricks. Wooden framed mirrors with concealed red neon accents expand the space with reflections and indirect views. The color palette for the loose furniture's upholstery is inspired to the "peacock's feather".

Raw natural materials, wood or stone floor finishes and red bricks walls, are combined with precious custom-made surfaces used as highlights: special designed back-lit lattice panels, a combination of dark coated metal and colored acrylics, provide a translucent screening to the 4 windows at the background and to the bottle shelves behind the bar counter. The depth of the elements defining this pattern, yet another variation of the peacock's feather recurrent theme, allows a multiplicity of transparencies and shades changing according to the point of view.

The ceiling is a composition of interlocked custom-designed metallic modules, vibrating with reflections. It is completely accessible for maintenance, and integrates installations, sprinklers, lights, and audio-video devices.

作为鸡尾酒吧，Brownstone 是一家夜场新贵，位于上海永嘉路的永嘉庭。它是与“Blue Frog 蓝蛙集团”以创新理念合力打造的餐饮项目。

酒吧的布局兼具简约与高效，当然，部分原因是由于室内空间有限。各种材质与专门设计的装饰元素相互交错、彼此映衬，共同营造了酒吧的高雅氛围，令人印象深刻。这里有20世纪初法国餐馆随性风格的沙龙，可以任由顾客享受悠闲时光。

酒吧在 120 平方米的空间内安放了两种不同高度和类型的座椅共 80 个，可以根据不同需要以不同密度重新布局、摆放。吧台处的两个平行鸡尾酒台可以同时提供服务。另外，还有一个 DJ 调音台、一个小型后厨及存储间。

酒吧中央区域围绕着吧台和 DJ 调音台，配有高脚凳和吧椅。周围有木制内嵌式柜台，与吧台同高，客人可由此进出抬升的外围平台休闲区。酒吧地板由四种石料制成的瓷砖铺成马赛克图案，明暗交错，在自然光下动感十足。

抬升的休闲区域的座椅高度较低，但舒适有加。共设有六组皮制的扶手椅和沙发，每组都配置了一个矮脚石面咖啡桌。

酒吧地板和固定家具都采用暗色木料，墙壁由红砖铺面。内置红色霓虹灯照明的木制镶框镜面通过反射扩大了空间感和视角。可移动装潢家具的颜色搭配则是从“雀翎”的图案中汲取的灵感。

天然的材质、木制或石制地板装饰和红色砖墙，以贵重的定做表面装饰来画龙点睛：由黑色涂层金属和彩色亚克力材料制作的背光隔栅系特别设计，以半透明方式屏蔽了后面的四扇窗子和吧台后部的酒瓶架。各种设计元素都服务于这种模式，增加了景深。然而，“雀翎”主题的不断重复变换，使得酒吧内部空间的各种透明质感与色调组合拥有了移步换景的神奇效果。

酒吧的天花板由专门设计的相互扣连的金属板材组成，具有动感的反光效果。天花板的维修极其方便，并连接着洒水器、灯饰和视听装置等各种设施。

Paul Kelly Design

The Polo Lounge

Designer
Paul Kelly

Paul Kelly Design, an award winning star brand boutique bar design company, was approached to design the 4-storey Oxford Hotel in 2005. The clients wanted to attract 2 separate markets in the development with the basement and ground floor targeted to the Sydney late night party market, and the top two levels as the ultimate VIP destination "Will and Tobys". Paul Kelly and the owner of the hotel Jason Gavin then went to New York, Las Vegas and Los Angeles to research the top VIP bars in these cities with over 200 bars and clubs visited, with the best of these bars conceptually used to develop the Will and Tobys' Spaces.

Level 1 and 2 are being operated by beverage brand Will and Tobys, a team synonymous with late night Sydney Style. The tone is of one of exclusivity, and this is apparent from the moment you enter off Oxford St. As you scale the glittering heights of the golden glow of the clear glass staircase that links both levels to the street, you are the ultimate VIP, entering a world of glamour, sophistication and old world charm.

At the zenith of the stair, is "The Polo Lounge", the ultimate Destination VIP Bar. The space is timelessly elegant, reminiscent of a fabulous old world Beverly Hills Hotel. The light levels are low, the people beautiful and your expectations are high. The pallet is kept to a bear minimum of chic black and white, with the walls clad in highly ornate white gloss timber panels with black marble trims. Beneath your feet is a beautiful hand made rug set into oak parquetry floor with classic leather buttoned chesterfields throughout. Finally, with the views across the city framed out by crystal glass and lead light trims of the balcony the tone has been set for the debut of a Sydney Hospitality icon.

The Polo Lounge's combination of striking black and white interiors and luxury lounge atmosphere are inspired by the timeless appeal of the gentlemen's clubs of London's Mayfair and New York's Upper East Side. The Polo Lounge was created as an escape from the maddening crowds of Taylor Square below, its late night, its Sydney and it's a very special place to be.

Will and Tobys has become an icon in the International Bar Scene, it has been nominated for several awards throughout the world, and has recently won the Top Venue design in Australia for 2007 for Paul Kelly Designs work. If you are in town, drop in, you won't be disappointed.

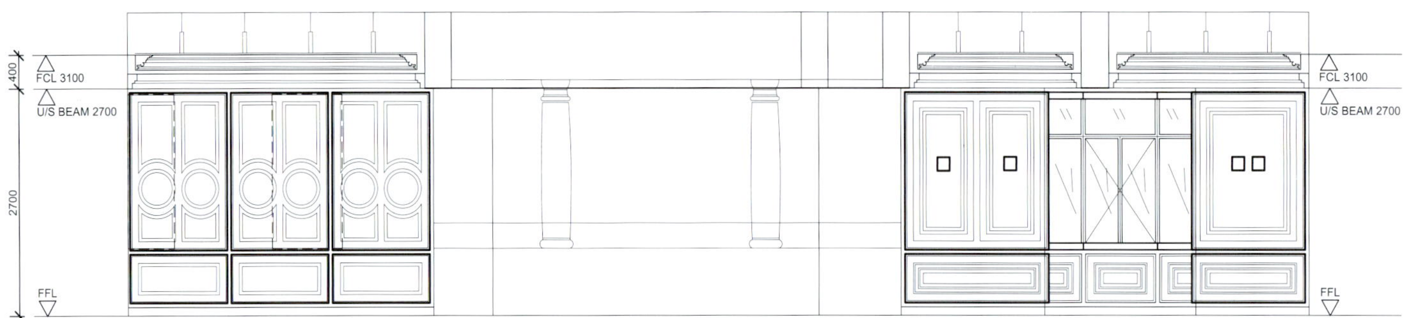

POLO LOUNGE - OXFORD HOTEL

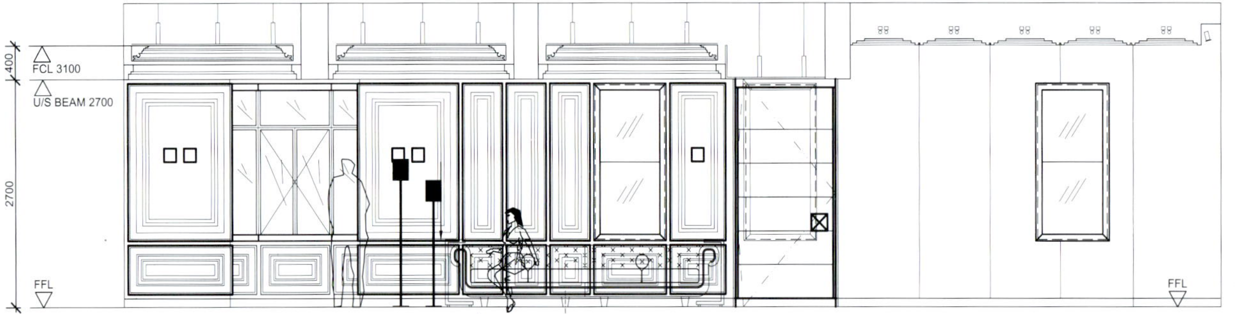

POLO LOUNGE - OXFORD HOTEL

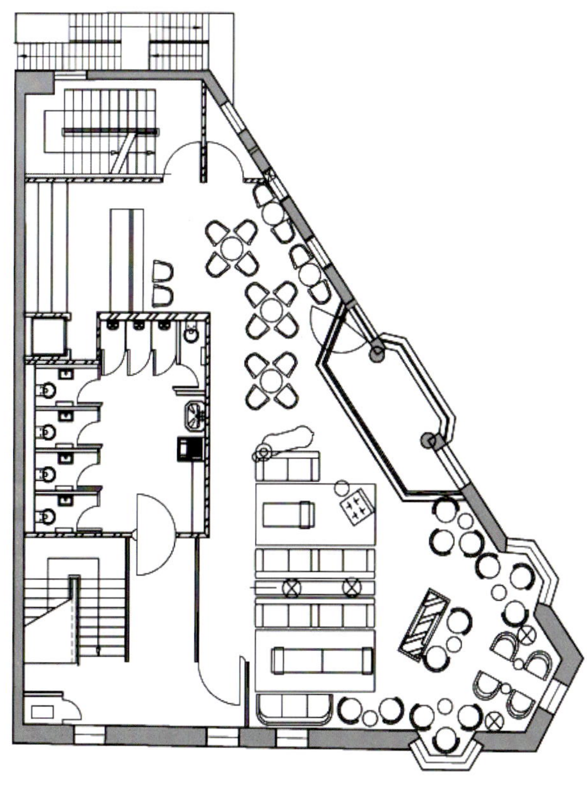

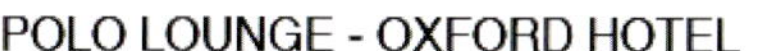
POLO LOUNGE - OXFORD HOTEL

Paul Kelly Design，一个屡获星级精品酒店设计奖项的设计公司，在2005年受邀设计了这座四层的万富酒店。客户希望建成的项目能够吸引两方面的客户群：酒吧地下层和一层面向悉尼繁荣的夜市市场，而上面两层则是为顶级VIP客户设计的酒吧“Will and Tobys”。设计师与这座酒店的开发商Jason Gavin曾赴纽约、拉斯维加斯和洛杉矶实地考察了200多个顶级酒吧，借鉴其最具特色的设计元素，最后打造出这座Will and Tobys。

酒店其中的两层由饮品品牌Will and Tobys经营，是悉尼夜生活的代名词。Will and Tobys强调尊崇享受，这一点，从你踏入牛津街的那一刻就能体会到。当你环顾连接着酒吧和地面的金碧辉煌的透明玻璃楼梯的时候，你就已经是这里的VIP，并且进入了一个迷人、精致而又散发着古典魅力的世界。

楼梯的尽头是最终目的地：贵宾酒吧“The Polo Lounge”。永不过时的优雅空间使人想起上世纪神话般的比佛利山庄大

酒店。低亮度照明和漂亮的人们使你对此处的期待值越发得高。整体色调较为单一，以时尚的黑色和白色为主，墙壁由黑色大理石装饰的带有白色光泽的木板所覆盖。在你的脚下是一张美丽的手工地毯，铺在橡木地板上，周边还有经典的纽扣式皮质长沙发。最后，水晶玻璃所呈现的全城景观和阳台的导光装饰奠定了悉尼图标酒店的处子秀。

The Polo Lounge酒吧将突出的黑白双色室内空间与豪华的休闲娱乐氛围相结合，其设计灵感来自于伦敦上流社会和纽约上东区对绅士俱乐部的无尽追求。The Polo Lounge酒吧专为那些想要从下面泰勒广场狂热的人群中逃离出来的人们所设计，其午夜经营和悉尼特性，使得这家酒吧成为人们的一个特殊去处。

Will and Tobys已经成为国际酒吧领域的代表之一，曾被世界各地的奖项提名且近期还为Paul Kelly Design赢得了2007澳大利亚顶级场馆设计的奖项。如果你在城里，就来这间酒吧吧，绝对不虚此行。

Designer
Aiji Inoue

Location
Tokyo, Japan

Area
150.55 m²

Photographer
Satoru Umetsu

This members' bar lounge targets to the executives who are keen in spending stylish city nights.

It goes without saying that you will feel something special and extraordinary for those who want to become a member and who are willing to use this bar. What's more, once you start using here, usability and comfort will play a great significance to the operation form.

Indeed, this "two aspects" was the keyword for designing this bar lounge.

Take one step into the reception hall. Warm lights that transmit the stones illuminate customers.

They find themselves reflected in the black-frame-mirrored wall surface, in which they have never seen before. Just by walking through this hall, it has a helpful effect of switching customers' moods.

Bookshelves occupying the walls look like a luxurious study room neighboring the hotel lounge. From this style, customers will feel the relaxed atmosphere. On the other hand, to make a contrast with the artifacts (books), we set natural rock face as reliefs to offer extraordinary impressions. Furthermore, we chose a luxurious color, champagne gold, to make good balance. We set a custom-made wine cellar (you can see inside from both sides) for dividing the spacious room. We made a see-through boundary to connect the modern space composition and authentic space composition. As for the VIP room tables, our image was to display moments of randomly-cut-wood into a glass case. Woods have the sense of stability and glasses have the brittle impression. When inverting these two designs, we can bring out two aspects of the materials.

Thus, we gave bilateral character to the lounge space and material, by controlling the constitution of the bar. As a result, we were able to create a sophisticated and high-grade bar lounge.

这家会员制酒吧致力于吸引那些热衷于时尚都市夜生活的消费群体。

对于想要成为这间酒吧的会员和想要体验这间酒吧服务的人来说，毫无疑问地，这间酒吧会让他们感觉特别和与众不同。更重要的是，一旦你踏进这里，可用性和舒适度会让你在这间酒吧中更得心应手。

事实上，可用性和舒适度这两方面是该酒吧设计的关键。

迈入接待大厅，温暖的灯光映照在小石头上，又反射到顾客的身上。

顾客会发现自己的影像倒映在装有黑色框架的镜面墙上，这是他们从未见过的。只要走过这个大厅，顾客的情绪就会得到转变，变得更为愉悦。

书架布满了整面墙，看起来就像是酒店大堂旁有一间豪华的书房。这样的风格设计会让顾客感到轻松。另一方面，为了与工艺品（书）形成鲜明对比，设计师将自然岩石面做成浮雕，给人非凡的印象。此外，设计还以奢华的颜色——香槟金，来保持空间的良性平衡；并定制了一个酒窖（从前后两面都可以看到酒窖的内部）来划分宽敞的房间。设计师制作了一个透明的边界以连接现代风格的空间和传统格调的空间。至于VIP房间的桌子，则是由随机砍下的木头放在玻璃橱里组合而成的。木头给人以沉稳的感觉而玻璃则给人以脆弱的感觉，将南辕北辙的两种元素用在一起，就会呈现出材料的两面性。

因此，通过控制酒吧的构造，设计在酒吧的空间和材料方面呈现出了两种不同的特点。最终营造出了一间精致、高档的酒吧。

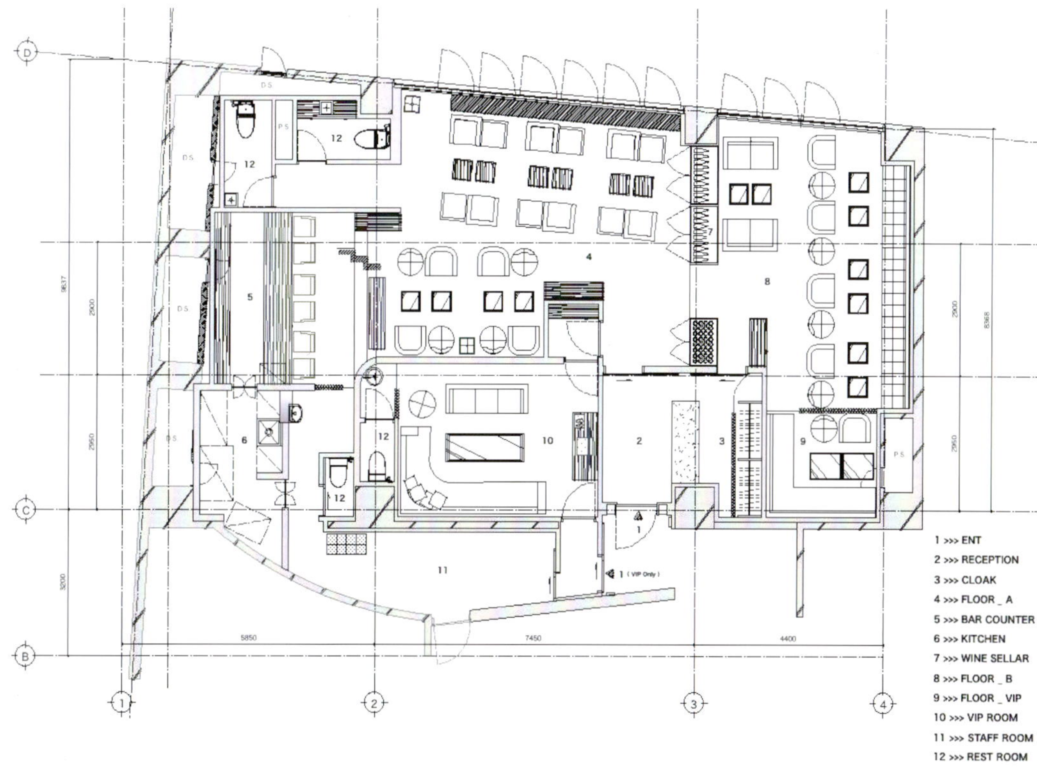

OOBIQ Architects

Designer
Giambattista Burdo, Samuele Martelli

Location
Shanghai, China

Area
400 m²

The building where Drop is located is one of the few buildings in downtown Shanghai with such original decoration on the ceiling and on the walls. It used to have its own character and soul.

With the definition of "contemporary classic" OOBIQ Architects refers to an interior made of classic textures, traditional shapes and materials, but transformed and adapted to a club to be located in such an international city as Shanghai.

That's why the interior started to have shapes that remind of European classic interior.

The space is organized on different levels, with the VIP rooms and the seating areas located on platforms. In this way the space is not flat and different areas where you can see and you can be seen are created.

The whole club is presenting the same floor finishing all around the areas – a geometrical Moroccan tile became the motif of the precious marble floor.

This warm decoration becomes another element that shows clearly the crossover of the interior design of Drop. Elements of different cultures and times connected together in order to create a unique atmosphere that has to be sophisticated and easy going at the same time.

The VIP room is created as a box where you can see all the interior space, and the design is based on the contrast of soft and hard material. Both the hard material outside and the soft material inside are decorated with the same pattern.

On the wooden walls, two installations custom-made for Drop make the environment more unique: some famous European paintings are "pixelized" and made three-dimensional as in the restrooms.

Drop is designed in order to provide an idea of intimacy but with a feeling of luxury, sophistication and grandeur.

Drop酒吧所处的建筑位于上海市区少数几个天花板和墙壁仍保持原有建筑风格的楼群之中。这些建筑均拥有自己的特色与灵魂。

欧比可建筑师事务所对于“当代经典”的定义是指，在室内设计中使用经典的纹理、传统的形状和材料，但对其加以修改以满足上海这样的国际都市的要求。

这就是室内设计能够使人们联想起欧洲经典室内设计的原因。

空间的布局具有层次感，贵宾室和坐席区位于平台上。这种布局呈现出的是一个立体的空间，客人既可以独自观赏四周，又可以变成被观赏的对象。

整个酒吧区铺设着统一的地面材质——一种几何形状的摩洛哥瓷砖，在稀有的大理石地面上拼出图案。这个温暖的装饰成为了表现Drop酒吧内部设计风格转变的另一要素。不同文化和时代的元素汇聚一体，营造出一个独特的氛围，使优雅与亲密感并存。

贵宾室的设计宛如一个盒子，置身其中，内部空间一览无余。软质材料和硬质材料的运用形成了鲜明的对比。无论是外部的硬质材料还是内部的软质材料，它们的装饰模式都是相同的。

木质墙上两处专门为Drop酒吧定做的装饰使其更加独特，一些欧洲名画被“像素化”，使其更具三维效果，洗手间也采用了这种设计风格。

Drop酒吧旨在营造一种亲密理念，同时又不失奢华、优雅与庄重之感。

Crox International Co., Ltd.

Shanghai La Lé Wine Bar

Designer
Tsung-Jen Lin

Area
$170\ m^2$

Location
66 Yandang Rd, Shanghai, China

Photographer
Shen Qiang

La Lé Wine Bar is a place where busy-urban people share casual evenings with friends under comfortable ambience enhanced with fine wines.

Located at Yandang Road, the earliest pedestrian street in Shanghai, La Lé Wine Bar is named with the pronunciation of the word "chatting" in Minnan dialect to highlight the idea of the place, to chat and to have good time with friends. The bar features an atmosphere of warm and lay back for wine appreciation in Shanghai metropolitan.

Evening Charm

The bar's black facade with fine metal curves making it a charm in the neighborhood. The sunset or moon like store sign visually loosen up tight shoulders. The relaxing vibes slows down pedestrians' footstep.

Gathering and Sharing

Responding to the close-by Fuxing Park where numeral Shanghai identical phoenix trees are planned, the interior begins from a trunk of a tree behind the entrance spreading out to space. It is to create the atmosphere where trees are the center of Chinese neighborhood for gathering and conversations underneath. A bar is a field for interpersonal communication. The design internperates time and space providing city new comers a connection of urben experience and local communities that will further generage a new drinking culture. Customized furniture and light selection are arranged to suit demands in different areas namely bar area, wine tasting area and private rooms.

Modern and Surreal

The bar's aesthetic is a blend of modern and surrealism, articulated through the use of hard-edged materials like marble and glass. The curvy ceiling contour, wooden wall melody pattern and polished stainless steel create the dizzy visual for a convivial vinous space. Customized wine cellar with constant temperature and humidity control enable large collection of wines remains in good quality. Large glass viewing window makes it the center piece in the bar.

East-West Culture Exchange

La Lé Wine Bar reflects local culture in wine sharing and storytelling. It is where smooth Western wine accommodates high-pressure modern Chinese and their memory of old times.

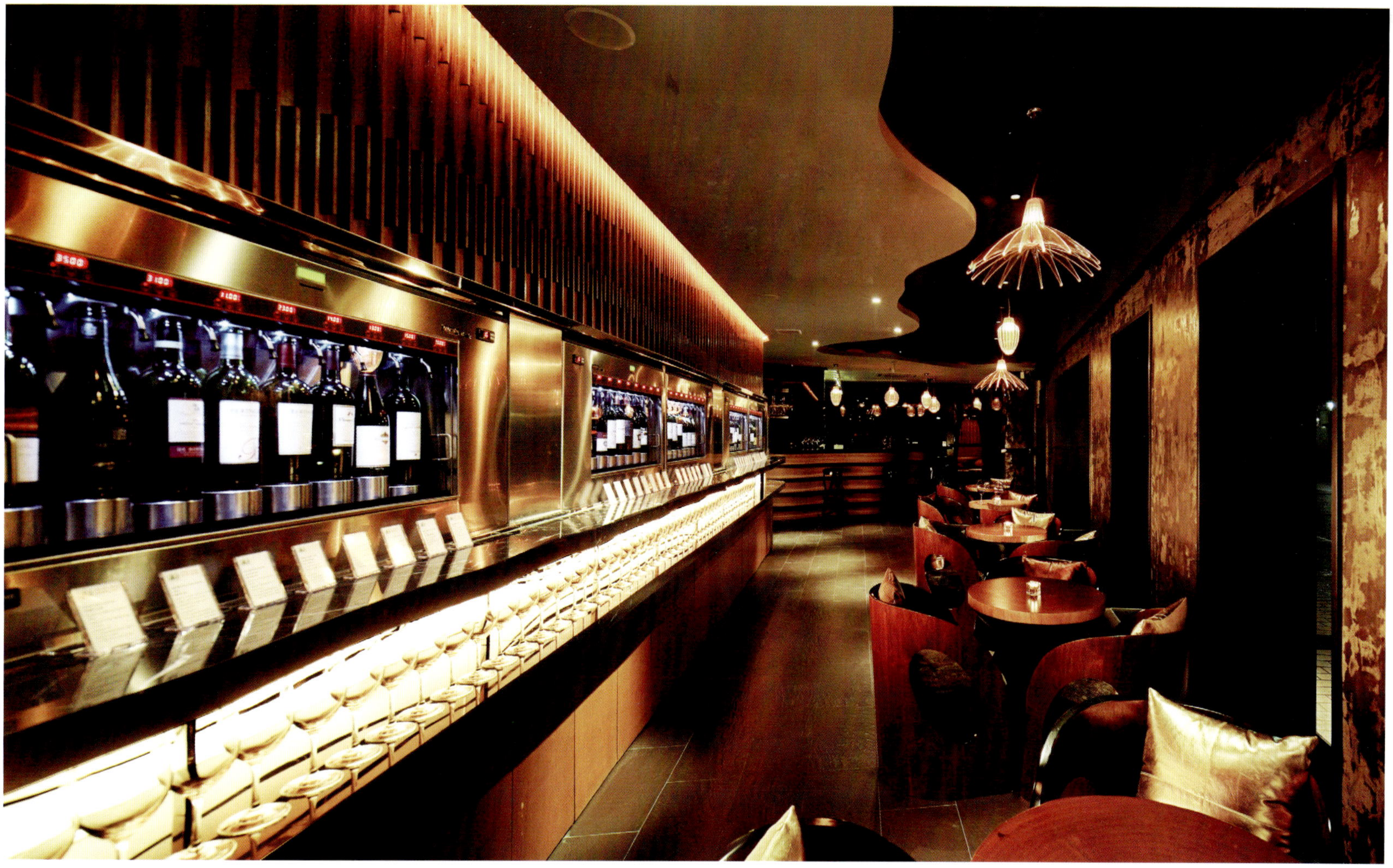

LaLé
LaLé wine bar
opening hours
4pm-2am

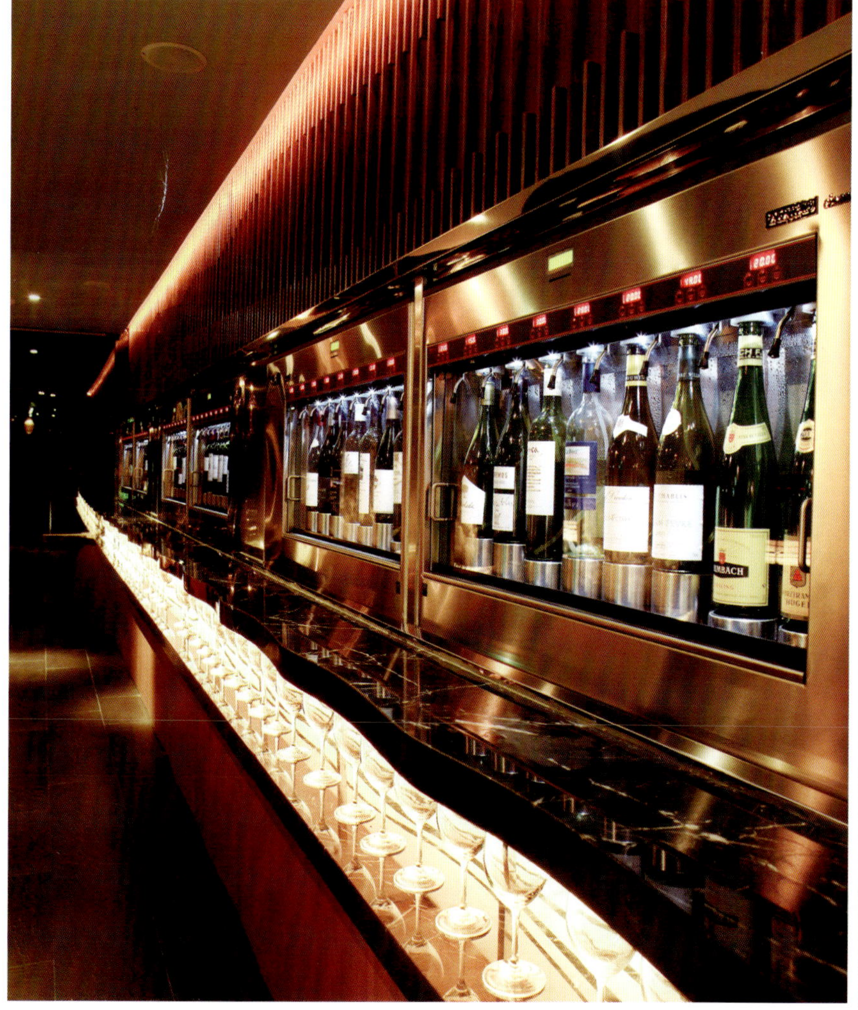

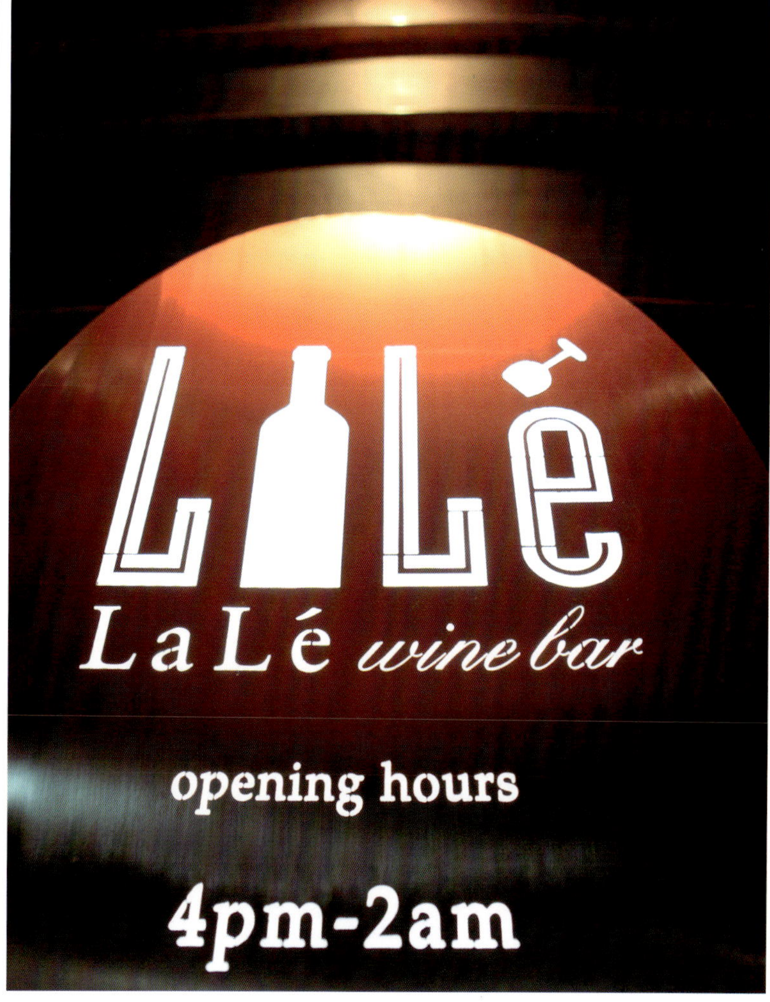

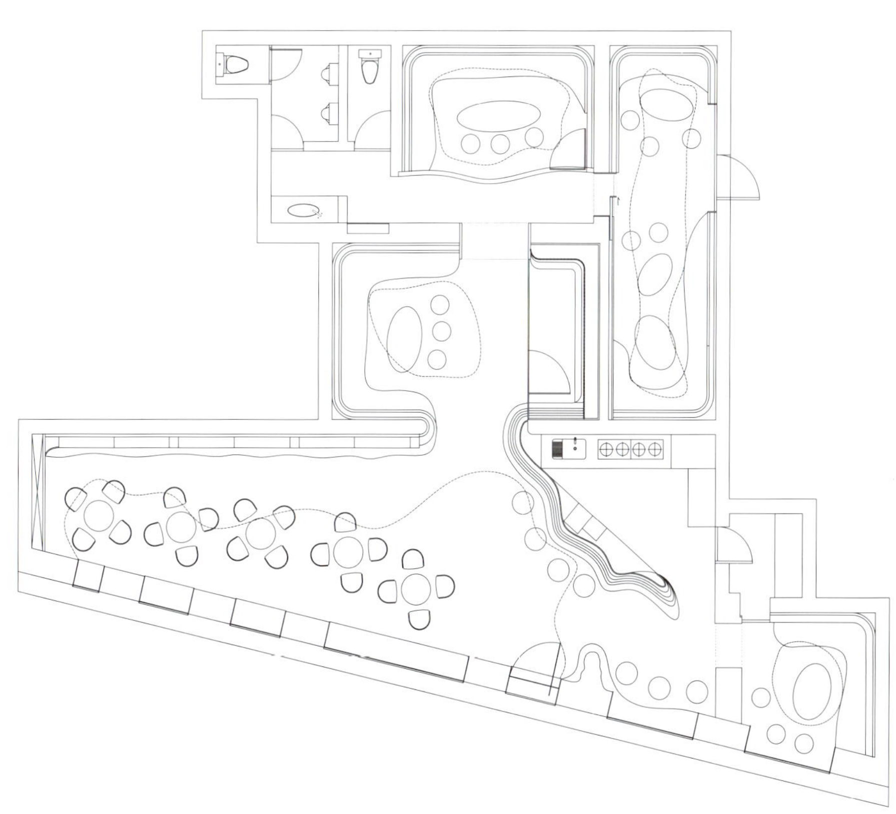

PLAN S=1/100

在La Lé红酒吧里面，繁忙的都市人和朋友们享受着惬意的夜生活，由葡萄美酒做伴，酒吧的氛围舒适宜人。

La Lé红酒吧开设在“雁荡路”上，这里是上海最早的步行休闲街。它的名字是由台湾方言中“闲聊”这个词的发音得名而来，意在凸显酒吧的经营理念：和好友聚会谈心，共享愉快时光。酒吧氛围宜人，实属上海这个大都市的品酒胜地。

夜幕中的魅力

酒吧的黑色外墙上饰有精致的金属曲线，彰显其魅力所在。如日落或如月亮般的店家招牌在视觉上缓解了人们的压力。轻松的氛围让行人放慢了匆匆的脚步。

聚会与分享

毗邻的复兴公园栽种着大量的上海梧桐树，La Lé红酒吧的设计与之遥相呼应。推门而入，源于一个树干结构设计的内部空间，一直延伸到整个空间。这是为了体现一种理念，即树下阴凉之处即是中国邻里之间聚会与言欢的主要场所。酒吧是一个人际交流的场所。这样的设计对时空进行着诠释，城市新人身处其中，可以体验到一种都市生活经历和地方社会生活的交融，进而产生一种新的饮酒文化理念。定制的家具与灯具选择同时满足了不同区域人们的品位要求，酒吧也相应地被分为吧台区、饮台区和独立包厢。

现代与超现实

酒吧的美感在于通过像理石和玻璃这样的硬边材料的使用，实现了现代元素与超现实主义元素的融合，进而满足了人们对高品位的期望。流线动感的天花板构造，如音符般排列出美妙韵律的墙面木材和精致钢材的运用为这个饮酒交流的空间营造出一份视觉上的虚幻感。量身定做的酒窖配有恒定的温度与湿度，使得在此库存的大量葡萄酒品质优良。大型玻璃可视窗的设置使其成为酒吧的中心区域。

中西文化交融

La Lé红酒吧在品酒和传诵故事中体现着本土文化元素。在这里，香醇的西方葡萄酒缓释着现代中国人的重重压力并唤起他们对悠久历史的回忆。

Yusaku Kaneshiro+Zokei-Syudan Co.,Ltd

"SOAN", the Japanese-style lounge has been born again as it was replaced. Counter tables that are made from tempered glass have been transplanted from previous place too, because they are still symbolic of the lounge. By making the walls filled with flower-patterned frames and uneven surface, the atmosphere was variegated.

Also effects of mirrors and indirect lights create more depth optically, and turning all color tones in the lounge down produces adult mood.

Designer
Yusaku Kaneshiro, Hiromi Sato

Location
Shinbashi, Tokyo, Japan

Area
38 m^2

Photographer
Masahiro Ishibashi

"SOAN"，这家日式酒吧以崭新的面貌重新开业。由钢化玻璃制成的柜台桌子是从先前的店里搬过来的，因为它们是这家酒吧的象征。墙壁上布满了花朵图案的相框装饰，凹凸不平，使这里的气氛也变得斑驳多彩起来。

同时，镜面效果和间接光源的应用还营造出视觉上的纵深感，使酒吧整体色调偏暗，带给人充满情调的氛围。

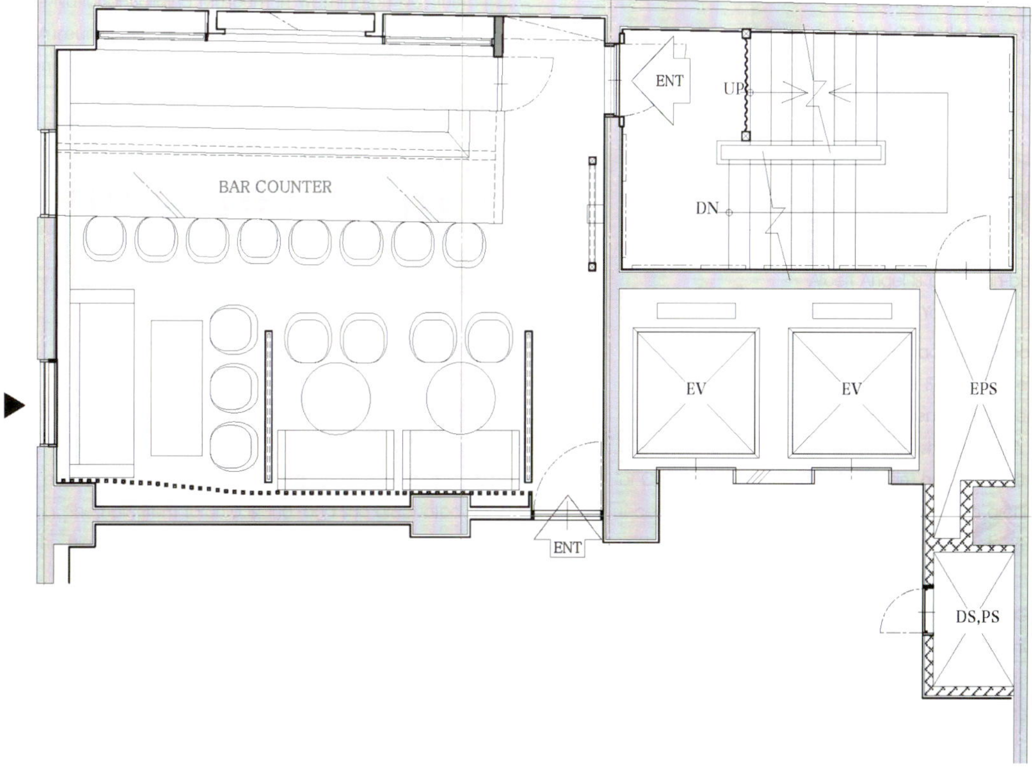
ENT
UP
BAR COUNTER
DN
EV
EV
EPS
ENT
DS,PS

草庵
草庵

G+ Design

Designer
Gulla Jonsdottir, Erni Taslim

Area
650 m^2

Location
Los Angeles, California, USA

Photographer
Edward Duarte

A psychedelic poetry garden was the first thing that came to mind when asked to design this project.

The guest enters through two 6.10 meters tall steel entry doors which reflect what goes on inside with its exploded cosmic flower, shapes of paint splashes and words of wisdom act as petals of the center flower shape, all custom-made in 3 different bronze and black color shades.

The outdoor patio walls are filled with 6.10 meters tall vertical gardens on both sides, with cut out steel tree and chain chandeliers and a custom-made hanging chair which sits high at the end where the lady of this poetry garden welcomes you.

The gardens from the patio move to the inside space in the transformation of a wall of black leather roses and the existing columns have been wrapped in laser cut steel sheets with carefully chosen words of poetry from such writers as Shakespeare, Edgar Allen Poe and E.E. Cummins. As the columns are backlit, the words project onto the floor and ceiling wrapping guests into this world of poetry. The existing brick walls received a layer of metal mesh and backlit poetry boxes.

The center bolder bar has a lotus flower sculpture above it with a dancer who performs each night and 5.49 meters tall ceilings contoured like a white cloud above you.

The VIP mezzanine structure looks over the interior space where guests can watch from above the leather custom-made booths that wrap around the space and the glass and mirror infinity cocktail tables with neon words all chosen from Elvis Presley songs such as: "Love me tender".

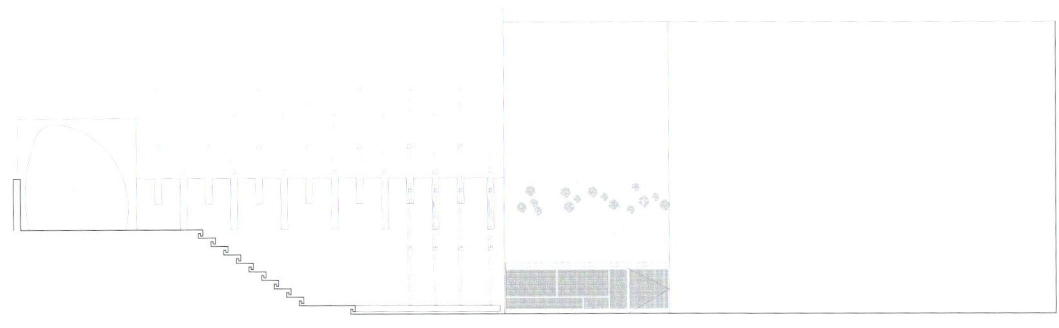

ELEVATION AT PATIO 1

SCALE: 3/16"=1'-0"

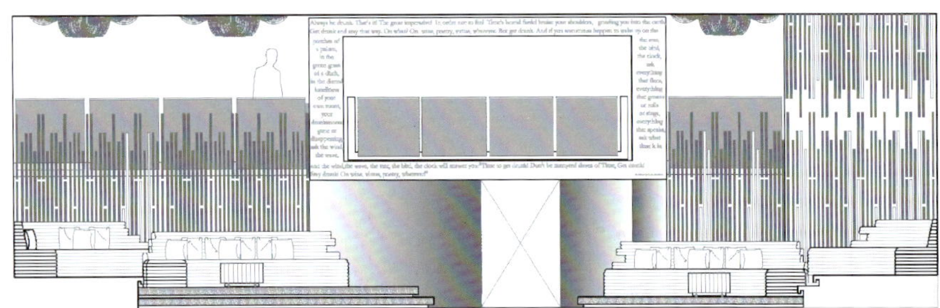

ELEVATION OF ENTRY HALL AND MEZZANINE 1

SCALE: 3/16"=1'-0"

受邀设计本案时，设计师首先得到的灵感是梦幻的诗意花园。

客人要穿越两扇钢制巨门才能进入内部，门高6.10米，巨大的爆炸状花形图案映射出内部的情景。喷溅图案和箴言妙句组成了花形图案的花瓣，所有这些都使用了特制的3种各异的青铜色和黑色。

酒吧外天井间的6.10米高的装饰墙的两面满布了园艺植物，还有铁艺树形装饰和连排吊灯。高处的定制吊椅上坐着诗意花园的服务员，欢迎着客人的到来。

天井的园艺墙延伸到了酒吧的内部空间，变成了黑色皮革制成的玫瑰装饰的墙面。原来的装饰柱由薄钢板包覆，上面用激光刻制了莎士比亚、埃德加·艾伦·坡和E.E.肯明斯等作家的名句。柱体由背光照明，这些诗句投射在地板和天花板上，将顾客包围在一个诗的世界中。原来的砖墙也装上了一层金属网格和饰有诗句的灯箱。

中央吧台十分醒目，上面有莲花雕塑，每晚都有一位舞蹈演员翩翩起舞。5.49米高的天花板被装饰得宛如头上飘着的一朵白云。

贵宾区的跃层结构可俯瞰酒吧，顾客可以在此居高临下地观看一个个由皮革装饰的隔间。玻璃和镜饰相互映照出无穷无尽的鸡尾酒桌，上面映出埃维斯·普里斯利（猫王）的名歌《温柔地爱我》等霓虹字样。

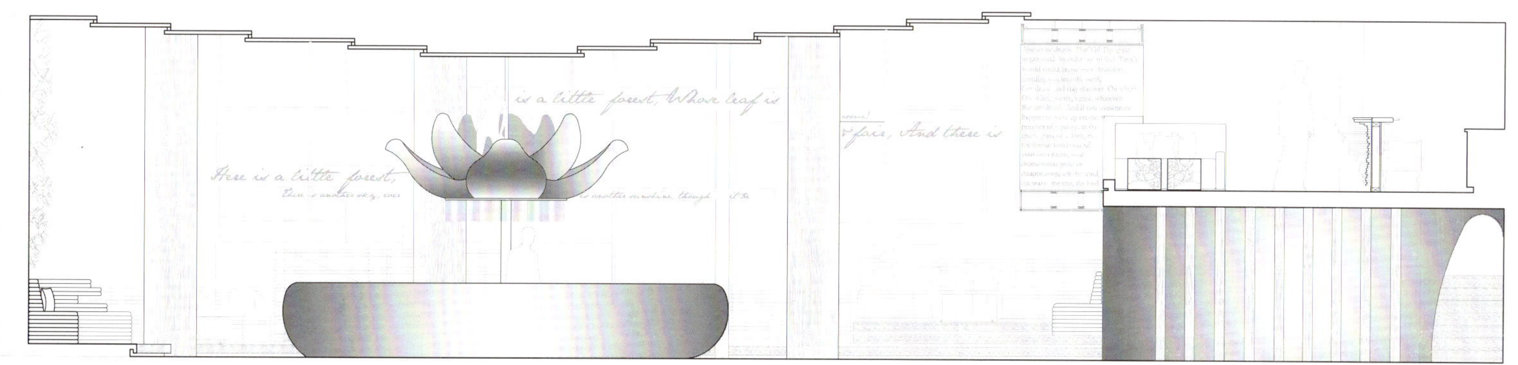

SECTION ELEVATION THROUGH BAR AND MEZZANINE 1

SCALE: 3/16"=1'-0"

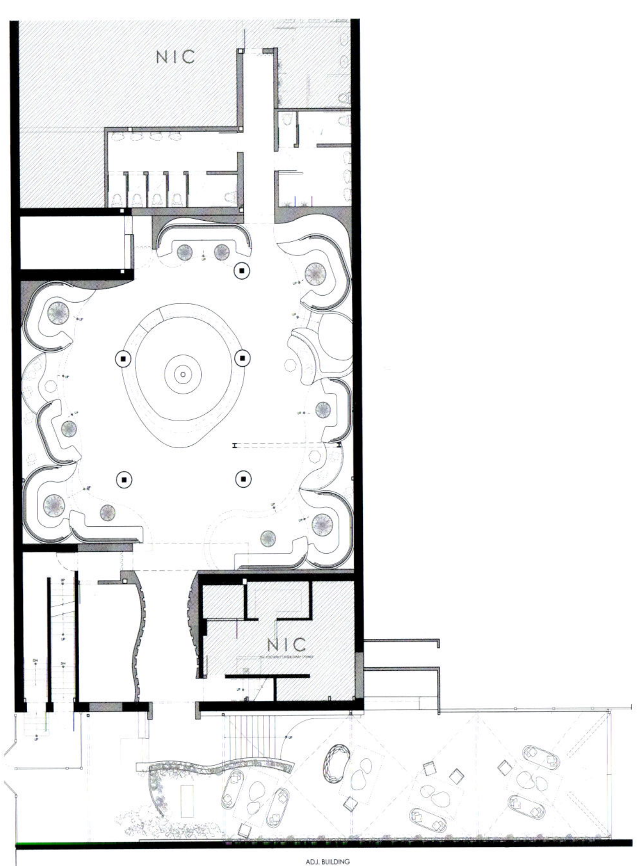

FIRST FLOOR PLAN 1
SCALE: 3/32"=1'-0"

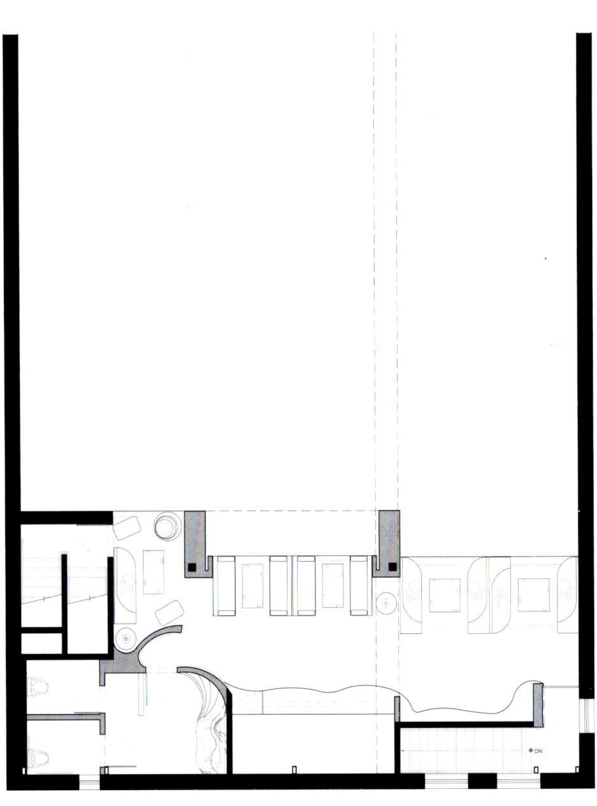

MEZZANINE FLOOR PLAN 1
SCALE: 3/32"=1'-0"

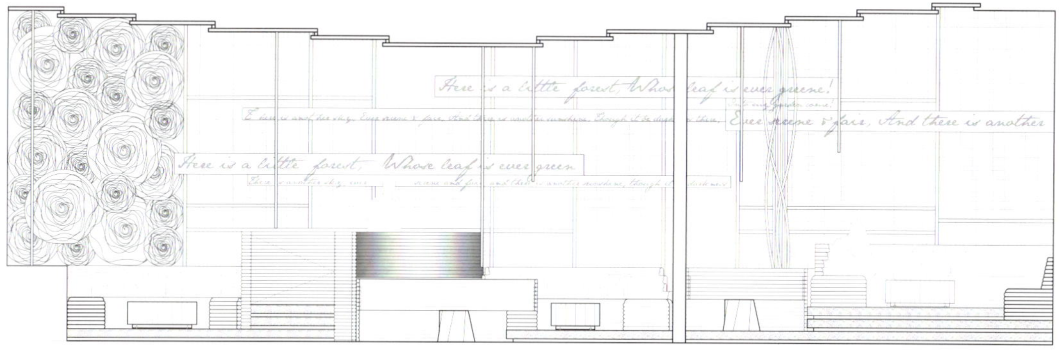

ELEVATION AT DJ AND SEATING BOOTHS 1

SCALE: 3/16"=1'-0"

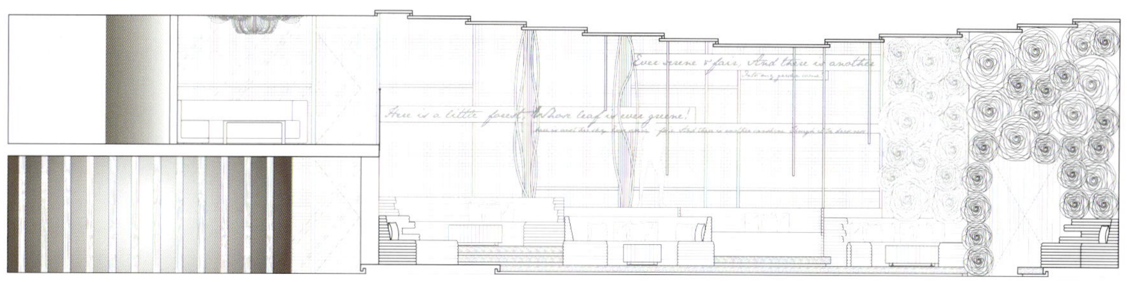

SECTION- ELEVATION THROUGH MEZZANINE AND SEATING BOOTHS 1

SCALE: 3/16"=1'-0"

Arch. FABIO FANTOLINO Studio

Big Club

Designer
Fabio Fantolino

Location
Torino, Italy

Photographer
Fabrizio Carraro

Trendy in Turin reality, well known by the inevitable city nightlife, the big club is a newly rebuilt with the ambition to become a reference point for the night life of Turin.

The Big aims in the guise of disco-lounge with refined contemporary rigor: the charm is guaranteed by the atmosphere and the soft neutral textures that change your mood thanks to the wise and mischievous use of lighting. The restaurant is located on two floors: the first sees the main body embraced by champagne-colored leather sofas and lights adjacent points, dominated by a front desk DJ golden marbles, precious and austere, dominated by the structure that makes velvet soul animation scene in the evening. The latter contrasts with the front bar area, which bar the fiberglass structure with soft lines is part of a scenario of glassy discipline.

The style is inviting and is wanted for the area around the main hall, surrounded by separate areas of the bottom floor, each in its own frame of marble slabs in the oblique cut and pleated curtains. The lines are decided, teams, spatial volumes that make the severity of a weapon of seduction. For further embellish the face of local areas are private room, lobby entertainment with their unique mirrored "sdiamantato" walls placed in the perimeter area.

The same logic of aesthetic taste is to reproduce the upper floor occupied by the restaurant "Metropole" which looks, with its chairs, right on the perimeter.

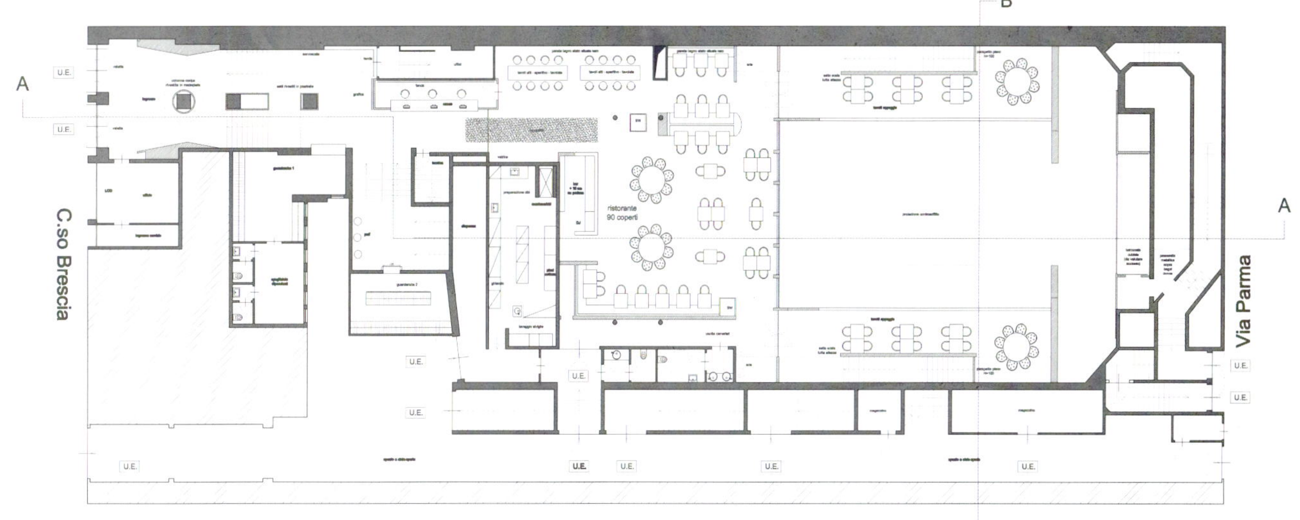

Big 俱乐部走在都灵的时尚前沿，在都市夜生活中闻名遐迩，重新装修以后的 Big 俱乐部致力于成为都灵夜生活的标志。

Big 俱乐部的目标是在迪斯科酒吧的名义下体现出现代文化中的精确性：通过灯光灵活巧妙的运用，室内的气氛和柔和的中性纹理能够调节你的情绪，进而保证了俱乐部的魅力。餐厅共有两层：第一眼便能看到室内的主要空间被香槟色的真皮沙发和交错相映的灯光所包围，灯光由金色大理石质前台中的 DJ 控制，体现出珍贵且简洁的风格，在夜晚，整个空间会因天鹅绒的使用而颇为灵动。后者与前面的酒吧区形成鲜明对比，在这里线条柔和的玻璃纤维结构成为了透明场景设计的一部分。

本案风格独具魅力，在主厅周围空间的设计上更为出彩。在一楼，主厅周围的空间各自独立，且由斜切的大理石板做柜架相互间隔开来，并搭配着打褶的窗帘。线条的设计果敢、有序、空间立体性强，因而成为极具魅力的武器。为进一步修饰，包间、娱乐大厅的局部区域配置的独特的镜像“sdiamantato”墙壁，也被放置在周边地区。

此设计与审美情趣重现了被“Metropole”餐厅占据的上层，因为“Metropole”餐厅的餐椅也放在周边，看起来与俱乐部风格很相配。

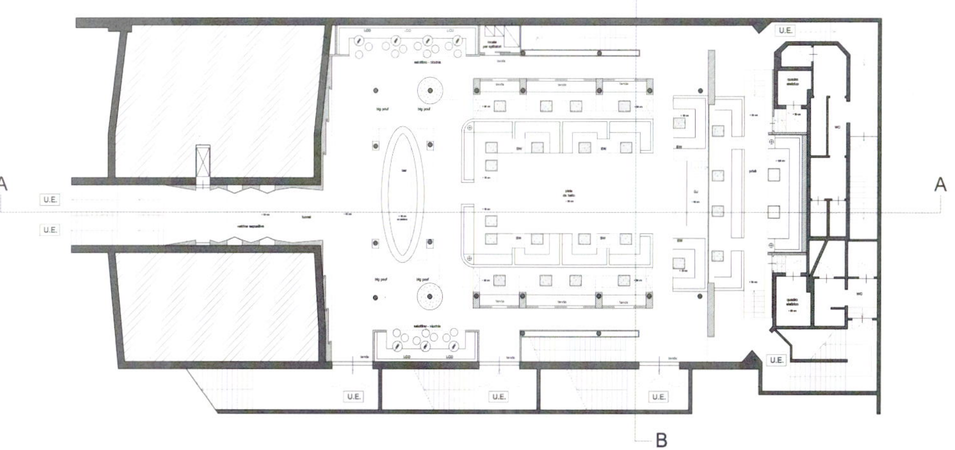

BIG

INDEX 索引

dwp

dwp is an award-winning, one-stop integrated design service, with global reach. Even in the most challenging of locations, over 450 multi-cultural professionals work together to deliver architecture, interior design, planning consultancy and project management, across borders, to the highest international standards. Divided into four distinct dwp portfolios, namely lifestyle, community, work and infrastructure, dwp integrated design services are managed and driven by the different dwp studios across the globe, ensuring dwp delivers each project with the qualities of a highly focused and specialized service, while offering diversity, flexibility and creativity over a broad spectrum. With currently 12 offices in 10 different countries, dwp presents its finest iconic designs time and again.

Virgile and Stone/Imagination

Virgile and Stone/Imagination is an integral part of the Imagination Group multi-disciplinary offer as specialists in Retail, Hotel and Leisure design. A London based practice of interior designers, architects and graphic designers founded in 1990, it is based in London West End within the Imagination building and has a presence within the offices where Imagination operates globally in the Americas and Asia Pacific.

In the 20 years since its creation it has gained an international reputation for design excellence, technological innovation and professional management. Fully committed to providing the highest level of service, it is able to bring together the most suitable resources to deal effectively with every aspect of a project, whether as lead consultant or as a member of a development team. Specializing in retail and leisure design, their services include Interior Design, Architecture and Graphics but also, through its insertion within the Imagination Group, many other complementary areas.

URBANTAINER Co., Ltd

URBANTAINER Co., Ltd is an architecture and design firm that creates unique atmospheres and spaces for subculture inhabitation within the city. Various social actions and behaviors are continuously taking place within the built urban environment. They seek this social activity and offer it a presence. Their designs are immediately inhabited by events for people to meet, interact, and enjoy. They give subculture a physical presence within the city. The urban condition in Asia is a complex vertical system of competition, architecturally and socially. They at URBANTAINER believe in a horizontal discovery and structuring of the urban environment; a social structure where unique communities can form and thrive. It is their philosophy to understand these communities and enhance them. Design does not only create physical beauty, but has the power to facilitate wellbeing and interaction.

Nikken Space Design

Originally established as the interior design department of NIKKEN SEKKEI LTD, Nikken Space Design (NSD) has more than 60 years of experience in the field. As a part of the Nikken group since 1994, NSD now works not only with Nikken Sekkei but is also in partnership with various architects and artists. With 60 highly experienced designers, NSD is currently working on numerous design projects for Japanese and international clients. NSD's wide range of projects cover interior design for hotels, high-rise offices, retail complexes, hospitals, educational institutions, restaurants, shops and even private residences. As a result, the company can accommodate all kinds of clients' needs. One of NSD's main principles is to bring high quality solutions in interior design, specifying furniture, fixture and equipment (FF&E).

ONG&ONG Pte Ltd

With a track record of more than 40 years in the industry, ONG&ONG offers a complete 360° solution that covers all aspects of the construction business. This three-pronged approach encompasses design (architecture, urban planning, interior, landscape, environmental branding, lighting and experience design), engineering (mechanical, electrical, civil, structural, fire safety and environmental) and management (project, development, construction, cost and place). They are an ISO14001 certified practice with offices in Singapore, China, Vietnam, India, Malaysia, the USA, Indonesia, Mongolia and the UAE.

Christina Zerva Architects

Christina Zerva Architects, founded in 1991, is a worldwide practice, working globally with project offices in Washington DC, Larissa and Belgrade. Christina Zerva leads the office with partners Paola Nena, Nastasia Sakic and Milan Li. The practice works internationally on residential, cultural and commercial projects. With each project they explore new ways to ensure that the best results are achieved through integration of concept and context with the programmatic and functional essence of a building. Through the continuity of the design process and extensive research involved in the analysis of site, program, social context, emerging construction and contemporary technologies, their approach and methodology introduce new perspectives and new dynamics, reinvigorating both landscape and cityscape.

Dariel Studio

Dariel Studio is a multi-award winning interior design company founded in Shanghai in 2006 by French Designer Thomas Dariel.

Since its establishment, Dariel Studio has completed over 60 projects of the highest quality in the main areas of design: hospitality, commercial and residential.

By confronting heritage and cutting-edge innovations, incorporating French design expertise with Eastern cultural influence, Dariel Studio's open approach unfolds new perspectives and forms while constantly observing clients' demanding standards.

The studio counts today a team of 25 professionals coming from various countries and backgrounds driven by the same passion for design.

Slade Architecture

Slade Architecture is an award winning New York firm founded in 2002 by Hayes Slade and James Slade. Their work has been recognized with international awards, publications and exhibits including Progressive Architecture Awards, NY AIA Awards, Award for Excellence in Design from the Public Design Commission of the City of New York, the Architectural League of New York Emerging Voices, Architectural Record Design Vanguard, Architectural League and Interior Design Magazine Best-of-the-Year Awards. Their work has been exhibited in la Biennale di Venezia, the Museum of Modern Art in New York, the Architectural League of New York, The National Building Museum in Washington DC, the Deutsches Architektur Museum in Frankfurt and many other international venues. They have been selected for the City of New York's DDC Design Excellence.

Lionel Ohayon & Siobhan Barry

Over the last 11 years, Lionel Ohayon, has grown ICRAVE from a two-person operation into a 35-employee, multi-million dollar business. A Canadian-born designer, Lionel Ohayon graduated from the University of Waterloo School of Architecture in 1994.

Siobhan Barry is a Partner and the Studio Director at ICRAVE and has been with the firm since its beginning. She hails from Toronto Island, and she graduated in 1997 and achieved the highest award (RAIC medal) for her graduating thesis in architecture.

She moved to New York after graduating, and established a career in retail/ boutique design and high-end residential. She then worked for MSM Architects (now MacKay Architecture/Design).

In 2002, Siobhan joined Lionel Ohayon to launch ICRAVE, a start-up design / build firm specializing in hospitality projects.

Kokaistudios

Kokaistudios is an award winning multi-disciplinary design firm founded in 2000 in Venice by Italian architects Filippo Gabbiani & Andrea Destefanis. Founded with the dream to create a collaborative office of young and talented architects devoted to researching and formulating the design solutions to the demands of tomorrow and capable of working on a worldwide basis; the firm has grown after 10 years into a team of 30 people headquartered in Shanghai, China.

With the focus of offering total design solutions, Kokaistudios has completed over 140 projects spanning the range of architecture, heritage renovation, and interior design. The firm has received numerous awards including the Asia Pacific Region of 2011 International Property Awards, Commercial Space in IAI Award 2010, MIPIM Asia Award 2010 and 2011, "40 under 40 Award" from Perspective Magazine. In addition to these awards Kokaistudios has also received 2 UNESCO Asia Pacific Heritage Awards for the Bund 18 and HuaiHai Lu 796 projects in Shanghai where Kokaistudios is responsible for the entire architectural and interior design projects.

Aiji Inoue

Aiji Inoue is CEO of Doyle Collection. Engaged as Interior Architect for 11 years at a Japanese famous design studio. Created many projects and received high evaluation in every project. Most significant features of his works are not just creating one scene but sequence. Just one year, Doyle Collection has been expanding the market from Japan to Asia.

Albert Angel

Albert Angel is an Italian-born architect and designer based in New York City.

From his nomadic childhood between Congo (DRC), Belgium and South Africa, he cultivates the taste of independence, the need for new horizons, the willingness to free oneself from established rules. A graduate of the University of Cape Town's architecture program he chose New York to begin his career, receiving training from some great names in architecture and design (Lindy Roy, Desgrippes Gobe, Ralph Appelbaum). In 2005, he founded his own architecture and design practice, Albert Angel.

His approach to design is directly inspired by the course of his life, far off the beaten track, each space, each object an invitation to a metaphorical journey, interior, dreamscape. The mixture of contemporary and classic influences, the marriage of design and art, are as much of the hyphens between each of his achievements.

TAI Architecture & Design Inc.

TAI HOSPITALITY DESIGN GROUP
ARCHITECTURE-DESIGN-FABRICATION-MANAGEMENT

TAI Architecture & Design Inc. is a Miami-based architecture and interior design studio specializing in the hospitality sector. The studio was born in 1998 by Francois Frossard, the lead designer. Since then TAI has designed some of the most recognized nightclubs and restaurants in the United States and abroad. It has also expanded its project roster to include retail, residential and hotel design.

Kryštof Blažek

2013 Interior of Aichisushi restaurant, Prague, Czech Republic

2013 Interior of Mustard agency office, Prague, Czech Republic

2012 Interior of Yaku restaurant, japanese grill, Prague, Czech Republic

2011 Interior of Café Restaurant Hřiště, Prague-Satalice, Czech Republic

2011 Interior of Bibitanuova restaurant, face-lift, Prague, Czech Republic

2010 Interior of Nana Vogue, vintage boutique, Prague, Czech Republic

2009 Interior of privat residencial apartment, furniture design, Prague-Žižkov, Czech Republic

2008 Interior of La Fuente, Music Club & Cocktails bar, Prague-Dejvice, Czech Republic

2004 Project of hair and beauty shop Gia, Prague Žižkov, Czech Republic

Alessandro Munge/Sai Leung

MUNGE LEUNG

Alessandro Munge

Managing Partner / Interior Designer

Alessandro has an uncanny ability to draw people together, and he eagerly feeds off the collective energy of the collaborative team. He pursues a dynamic approach to business and design, and has proven to have an exceptional aptitude for both. Taking the various intricacies of the client's mandate and distilling it into an exciting conceptual reality integrates two of his key strengths; his innate ability to capture the idea and the clarity in which he sees the goal both architecturally and financially.

Sai Leung

Design Director / Interior Designer

Understated by nature, Sai's refined design sensibility guides the studio's projects along with a fine hand, and perceptive eye. The consummate mentor, he directs the studio with meticulous attention, providing vision and leadership for all team members. Particularly gifted in the art of lighting design and custom furniture, Sai remains very hands-on in all aspects of the process. As concerned with the overall as intimate detail, he tenaciously pursues the project goals, and the results demonstrate conceptual clarity and keen sense of craft.

Francois Frossard

Francois opened his own design firm Francois Frossard Design (FFD) In 1998 in Miami. He pioneered a new era in nightlife design. He set new standards in interior design around the world.

Francois confronted each venue with a careful meditation on style, function and originality; a marriage of art and business with a profound understanding of his extremely fickle clientele. Francois deals with the complexities of building multiple venues in various cities at the same time masterfully, considerate of local building codes, politics, labor laws and community; designing showpieces capable of withstanding time. Respectfully, Francois cherishes each day as a gift and embraces tomorrow's mystery. He does not take his good fortune for granted and applauds his team, talented craftsmen, manufacturers, designers and tradesmen that service his projects state-wide and internationally. As his company expands further into world class markets his grounded classically trained past narrate his future.

Alexander Lervik

Born 1972 in Stockholm, Sweden

Alexander started his education in design as a carpenter apprentice for two years, and then studied interior design and product design at Beckmans School of Design for three years (1995~1998).

The graduation project from Beckmans School, the exhibition "Ten Stools – Ten Decades", has toured around the world for six years, and has been seen by over one million people.

Mainly, stores, nightclubs and offices have been designed. Some of the designs include:

Restaurant Supper Stockholm 2009

Restaurang Curman/sturebadet Stockholm 2009

The nightclub Push in Gothenburg, Sweden, January 2008

Elia Felices

Elia Felices was born in Almería (Spain) in 1973. Now she is defined as a scenic interior designer, creating environments which seek to generate visual impact on the "Viewer", evoking feelings of peace, relaxation, emotion... and all this in a simple and straightforward, staging it exempted overburdened elements. Elia's projects don't neglect the commercial purpose, making interior design a profitable investment for their customers, they view as their businesses acquire power on brand providing added value. Proof of this is projects such as Dalai disco, New Kuvee or Finques La Llar upgrades.

Archinexus

Archinexus, founded in 2004 by Mr. Tai-Lai Kan, a leading-edge design practice based in Taipei, China, is a creative and collaborative design firm with extensive practical experience, especially renowned for strong design across a variety of high-profile projects. Dedicated to innovative problem solving and design excellence, Archinexus is a team of highly qualified design professionals daring to dream as well as those who are capable of making dreams come true.

Paul Kelly

Paul has spent the last 15 years working within the industry and truly immersing himself into every detail of what makes the international entertainment industry come to life. Paul's passion is authenticity and creativity, this he believes is crucial to the success of his venues, of which he has created in excess of 150. Paul Kelly frequents the US, Europe and Asia to ensure his brand direction and the businesses he creates are not only international icons but lead the market with innovation, recognition and success. Paul Kelly's attention to detail and brand philosophy will ensure a seamless journey when developing your dream.

Mr. Important Design

Mr. Important Design is an interior design firm that brings an ebullient perspective to interiors. Specializing in hotels, nightclubs, restaurants, bars and lounges, Mister Important Design works closely with clients to achieve interiors that exceed expectations. Exuberant interiors that are designed to be remembered and talked about.

An enthusiastic embrace of many time periods, styles, intents and esthetics are sampled to create interiors the way hip-hop artists sample to create new music. Everything is available and usable, and the sparks fly best when kitsch and glamour, new and old, high and low all rub up against each other.

Collaboration with emerging designers and artists world-wide keeps the work fresh and surprising. Use of cutting edge technologies in lighting and materials coupled with a deep background in traditional furniture and decor help produce the unexpected thrill of these spaces.

The firms design work has been published widely internationally. Recent accolades include "Best Nightclub Interior, 2010" – Hospitality Design Magazine and "Best International Nightclub Interior, International Media Prize 2010" – Modern Decoration, China.

Concrete Architectural Associates

concrete Concrete builds identities, is a multi-disciplinary creative office, from urban planning to interior design, from architecture to graphic design.

Concrete originally was founded by Rob Wagemans, Gillian Schrofer and Eri van Dillen in 1997. They met each other by a not realized project, a head office in Amsterdam for Crque du Soleil.

Gillian Schroter left concrete in 2004 to start his own company.

Rob Wagemans is founder and creative director of concrete.

Lisa Hassanzadeh is head of interior design and partner of concrete.

Erikjan Vermeulen is head of architecture and partner of concrete.

Sofie Ruytenberg is head of identity, concept design, brand development, visual marketing and press and is associate of concrete.

Erik van Dillen is creative advisor.

Parolio & Euphoria Lab

Parolio & Euphoria Lab is an interior and communication design agency founded in 2003 by creative director Parolio and responsible for some of Spain´s hippest, more dynamic and exciting new spaces with six clubs and bars opened in the last two years, including classics like Pacha, it is inevitable that if you visit Spain you´ll end up experiencing some of Parolio & Euphoria Lab´s work. Parolio is also responsible for the naming, branding, image and creative direction of the projects.

The agency also offers international trend consulting and market research for companies worldwide, especially in Asia and Latin America.

Antonio Di Oronzo

Antonio Di Oronzo came to New York from Rome (Italy) in 1997 and has been practicing architecture and interior design for eighteen years. He is a Doctor of Architecture from the University of Rome "La Sapienza", and has a Master's in Urban Planning from City College of New York. He also holds a post-graduate degree in Construction Management from the Italian Army Academy.

In 2004, Antonio founded the award-winning firm bluarch architecture + interiors + urban planning, a practice dedicated to design innovation and technical excellence providing complete services in master planning, architecture and interior design.

At bluarch, architecture is design of the space that shelters passion and creativity. It is a formal and logical endeavor that addresses layered human needs. It is a narrative of complex systems which offer beauty and efficiency through tension and decoration.

Tsung-Jen Lin

Tsung-Jen Lin, Director of Crox International, specialized in architecture and urban planning, has a good sense in design and a sharp business understanding in all aspects of a success project. His practice spans from window display, scenery, interior, to large scale projects like trade show, exhibition and urban landscape. With unique design ability and user-focused approach, Mr. Lin looks into the context for solutions beyond traditional ways. He serves as a cohesive force that fosters an integrated design consulting service at Crox.

Education

Master of Urban Management and City Design, Domus Academy, Milano, Italy

Bachelor of Architecture and Urban Planning, Chung-Hua University, Taiwan, China.

Fred Mafra

Graduated in 1998 from PUC / MG being a member of Company M&SG – Architecture and Environment where he majored in Environmental Design, Architecture, Industrial and Corporate. Fred also devised all the design works of the University FEAD/ Minas among other projects such as lounges, restaurants, condos and night clubs in Belo Horizonte.

Gulla Jonsdottir

Critically Acclaimed Designer, Gulla Jonsdottir, a 2009 recipient of Hospitality Design's "Wave of the Future" award, is the creative and visionary force behind some of the most striking hotels, restaurants, nightclubs, and spas in the world. As founder of her new design firm, G+ Gulla Jonsdottir Design, located at La Peer Drive and Melrose Ave., Jonsdottir offers her clients a myriad of services including design, architecture, lighting, interiors, and furniture.

Icelandic-born Jonsdottir, spent eight years as Vice President and Principal Designer of the acclaimed Dodd Mitchell Design (DMD) firm in Hollywood where she was responsible for envisioning and designing numerous projects including the Hollywood Roosevelt Hotel, Cabo Azul Resort in Los Cabos, Mexico, The Thompson hotel in Beverly Hills among many others.

Arch. FABIO FANTOLINO Studio

In the professional panorama of Turin, Arch. FABIO FANTOLINO Studio is a well-established and a recognized point of reference for everything related to design and architectural design. Located in the heart of Turin, in a vein of the center, the location reflects the elements that identify the modus operandi and the creative and stylistic sensibilities of those who work there: spacious, exquisitely sophisticated, essentially refined, designed to accommodate customers equally ambitious in terms of exclusivity and refinement. A dynamic, attentive, receptive approach influenced by the advangarde reality of today and proactive in advancing the most cutting-edge ideas and solutions that meet customer requirements in terms of accuracy and satisfaction.

Orbit Design Studio

Orbit Design Studio is an award winning creative multi-disciplinary design agency with offices in Bangkok, Singapore and London. The company was established in 1996 and has developed a reputation internationally for innovative design solutions that are both memorable and successful. Their ability to integrate architecture, interior design and brand development from the beginning of the process allow them to deliver truly integrated projects that invariably enhance and build businesses and brands. Known for their hospitality, workplace, retail and brand designs internationally, Orbit works with a multitude of leading companies including BMW, Nike, PepsiCo, Diageo, Swatch, Louis Vuitton, Proctor & Gamble and the World Bank.

OOBIQ Architects

OOBIQ Architects is an award winning architectural and design firm based in Hong Kong and Guangzhou. OOBIQ Architects' founding partners are Italian architects, Giambattista Burdo and Samuele Martelli, graduates of the University of Florence who have also lectured, researched and supervised undergraduates of architecture.

After years of experience in European design firms working on projects all around the world, and after experienced in Hong Kong and Shanghai, Mr Burdo and Mr Martelli formed a partnership to combine their notable architectural skills, and brought Italian quality and style to Hong Kong, China and China mainland. Both of them are members of Unesco Forum.

OOBIQ Architects' design services include architecture, interior design, event and product design, artistic direction and project management. OOBIQ Architects' clients are both European and Chinese brands with high visibility who demand a quality product with personalised service.

Blacksheep

Blacksheep is in the business of creating signature spaces. Formed in 2002 by husband and wife team Tim Mutton and Jo Sampson, Blacksheep has a global reputation for the creation of exceptional hospitality environments, which perform both critically and commercially.

The studio boasts specialist teams who understand the importance of an all-encompassing, holistic hospitality design service including local market analysis, brand positioning, innovative concept creation/implementation and integrated graphics and branding.

As well as redefining London's nightlife scene over the past decade, smash hit nightclub, The Cuckoo Club, "Whisky Mist at Zeta" in the Hilton Park Lane. Other clients include projects for Four Seasons Hotels & Resorts, Starwood Hotels & Resorts, Accor Hotels and Jamie's Italians Restaurants.

Yusaku Kaneshiro+Zokei-Syudan Co., Ltd

Yusaku Kaneshiro

1960 Born in Okinawa

1979 Graduated from Okinawa prefectural Koza high school

1981 Graduated from Tokyo Designer Gakuin College

1988 Joint established Planning and Analysis for Environment

2000 Established Yusaku Kaneshiro + Zokei-Syudan Co., Ltd.

Hiromi Sato

1976 Born in Akita

1997 Graduated from Ogaki women's college Department of Art and design.

1999 To leave the PLANNING AND ANALYSIS FOR ENVIRONMENT.

2000 Joined Yusaku Kaneshiro + Zokei-Syudan Co., Ltd.

后记

本书的编写离不开各位设计师和摄影师的帮助，正是有了他们专业而负责的工作态度，才有了本书的顺利出版。参与本书的编写人员有：

URBANTAINER Co., Ltd, Orbit Design Co., Ltd., Elia Felices, Hayes Slade, James Slade, Christina Zerva Architects, Tsung-Jen Lin, Yusaku Kaneshiro, Hiromi Sato, Takeshi Nakano, Nikken Space Design, Nobuko Suzuki, Yudai Watanabe, Virgile and Stone / Imagination, Tomoyuki Hino, Fabio Fantolino, Francois Frossard, Aiji Inoue, dwp| design worldwide partnership, Thomas Dariel, Lynn Ng, Teo Boon Kiat, Jacques Hanekom, Albert Angel, Jonathan Legrand, ICRAVE, Antonio Di Oronzo, KAN, LAI, Lionel Ohayon, Siobhan Barry, Andrea Destefanis, Pietro Peyron, Carmen Lee, Paul Kelly, Giambattiata Burdo, Samuele Martelli, Gulla Jonsdottir, Erni Taslim, Paralio, Giancarlo Garofalo, Alessandro Munde, Sai Leung, Mehari Manna Seare, Fred Mafra, Alexander Lervik, Rob Wagemans, Onne Walsmit, Tim Mutton, Jo Sampson, Francois Frossard, Antonio Di Oronzo, Kryštof Blažek, Mr. Important Design, Sun Namgoong, Owen Raggett, Anton Stark, Mihajlo Savic, Shen Qiang, Shinichiro Sugai, Seigan Kuroda, Minoru Karamatsu, Fabrizio Carraro, Tom Arban, ADO, Ricardo Labougle, Jomar Graganca, Simon Hare, Satoru Umetsu, Jan KuděJ, Rafael Vargas, Masahiro Ishibashi,Courtesy of dwp, Derryck Menere, See Chee Keong.

ACKNOWLEDGEMENTS

We would like to thank everyone involved in the production of this book, especially all the artists, designers, architects and photographers for their kind permission to publish their works. We are also very grateful to many other people whose names do not appear on the credits but who provided assistance and support. We highly appreciate the contribution of images, ideas, and concepts and thank them for allowing their creativity to be shared with readers around the world.